SpringerBriefs in Microbiology

Guadalupe García-Elorriaga
Guillermo del Rey-Pineda

Practical and Laboratory Diagnosis of Tuberculosis

From Sputum Smear to Molecular Biology

 Springer

Guadalupe García-Elorriaga
National Medical Center La Raza, CMNR
Mexican Social Security Institute, IMSS
Mexico City, Mexico

Guillermo del Rey-Pineda
Department of Infectology
Federico Gomez Children's Hospital
Mexico City, Mexico

ISSN 2191-5385   ISSN 2191-5393   (electronic)
SpringerBriefs in Microbiology
ISBN 978-3-319-20477-2   ISBN 978-3-319-20478-9   (eBook)
DOI 10.1007/978-3-319-20478-9

Library of Congress Control Number: 2015944099

Springer Cham Heidelberg New York Dordrecht London

Springer International Publishing AG Switzerland is part of Springer Science+Business Media (www.springer.com)

Preface

This book is the result of a joint effort acknowledging the challenge of writing and publishing a book on the diagnosis of tuberculosis (TB). It is particularly appealing due to its advantage over other books, since it specifically focuses on the diagnosis of TB, encompassing the elemental diagnostic methods up to cutting-edge technology-based tests, including the diagnosis of TB infection (latent TB infection, LTBI).

This treaty is exclusively centered on the diagnosis of TB, including the spectrum of clinical diagnosis through the microbiological and molecular gold standard, the most practical, due to its celerity and high sensitivity. The diagnosis of LTBI, key to TB control, is also addressed.

Since TB diagnostic methods are still evolving, training must also be continuous. Great advances in this dynamic and ever-changing field have developed in the past few years, particularly resulting from the introduction of Molecular Biology. But unfortunately, this has led to increased costs and hence great disadvantages, leaving many patients without a timely diagnosis and appropriate treatment, particularly in highly endemic countries.

With comprehensive mastery, a change in the paradigm on TB diagnosis could well revitalize the required technology, making it more efficient, faster, predictable, and at a more accessible cost.

A century after Robert Koch discovered the bacillus causing TB, a great number of countries still depend on bacilloscopy as the only means of disease detection. We build on the past and we are all a product of our parents, professors, and colleagues as well as of our God-given talents and challenges.

I have had the privilege of working on the routine microbiological diagnosis of TB at the *Laboratorio de Microbiología de la Unidad Médica de Alta Especialidad* "Dr. Gaudencio González Garza," all the way to the Molecular Biology techniques in the Immunology and Infectious Disease Research Unit of the Infectious Disease Hospital at the *Centro Médico Nacional* "La Raza," affiliated with the *Instituto Mexicano del Seguro Social.*

We believe that particularly in countries with high TB endemicity, a quick and handy reference book on the diagnosis of TB is useful for Clinicians, Microbiologists, teachers and students of Medicine and Microbiology.

I wish to express my gratitude to many colleagues and physicians for their support and close collaboration, particularly during those fruitful meetings of the Center for National Epidemiological Surveillance and Disease Control (*Centro Nacional de Vigilancia Epidemiológica y Control de Enfermedades*). Also, to the *Dirección General de Epidemiología. SS*, primarily for the "Modification of the Mexican Official Policy NOM-006-SSA2-1993," for the prevention and control of tuberculosis at the primary health care level, published in the *Diario Oficial* on September 27, 2005; and secondly, for the elaboration of the "Practical guide to the care of tuberculosis in children and adolescents," in association with the National Tuberculosis Program, ISBN 970-721-334-5. December 2006.

I must also especially acknowledge all those silent heroes that have been of great assistance in the preparation of this manuscript: Gabriel Natan Pires, the Clinical Medicine Associate Editor that wholly believed in the Project; the always patient and kind Associate Editor at Life Sciences and Biomedicine at Springer Brazil, Roberta Gazzarolle Del Rossi, and our attentive project coordinator, Susan Westendorf.

My coauthor, Dr. Guillermo del Rey-Pineda, an expert Immunologist, and I hope that our initiative will motivate the interest of our readers not only in the solution of TB clinical diagnostic dilemmas but to prompt them to present new questions on routine and basic diagnosis, fostering a continuous bidirectional exchange between the realms of health care and those of clinical and basic research.

Mexico DF, Distrito Federal, Mexico   Guadalupe García-Elorriaga

Contents

Abbreviations

ADA	Adenosine deaminase
AFB	Acid fast bacteria
ATB	Active tuberculosis
ATD/GC/MS	Automated thermal desorption, gas chromatography, and mass spectroscopy
BAL	Bronchoalveolar lavage
BCA	Breath collection apparatus
BCG	Bacilo de Calmette-Guérin
CD1	Cluster of differentiation 1
CDC	Centers for Disease Control and Prevention (USA)
CFP-10	Culture filtrate protein 10
CPA	Cross priming amplification
CRI	Colorimetric redox indicator
CT	Computed tomography
CXR	Chest X-ray
DNA	Deoxyribonucleic acid
DST	Drug susceptibility testing
DTH	Delayed-type hypersensitivity
EPTB	Extrapulmonary tuberculosis
ESAT-6	Early secretory antigenic target 6
FDA	Food and Drug Administration (USA)
FIND	Foundation for Innovative New Diagnostics
HDA	Helicase-dependent amplification
HIV	Human immunodeficiency virus
HRCT	High resolution computed tomography
ICT	Immunochromatography test
IGRA	Interferon gamma release assay
IFNγ	Interferon gamma
IL-1b	Interleukin-1 beta
IL-12	Interleukin-12
IL-15	Interleukin-15

IL-18	Interleukin-18
ISTC	International Standards for Tuberculosis Care
LAMP	Loop-mediated amplification
LED	Light-emitting diode
LJ	Lowenstein–Jensen media. A solid culture media used for tuberculosis diagnosis
LPA	Line probe assay
LRP	Luciferase reporter phage
LTBI	Latent TB infection
MDCT	Multidetector-computed tomography
MDR-TB	Multidrug-resistant tuberculosis
MGIT	Mycobacteria Growth Indicator Tube
MODS	Microscopic-observation drug-susceptibility
MRI	Magnetic resonance imaging
MTB	*Mycobacterium tuberculosis*
MTBC	*Mycobacterium tuberculosis* complex. A genetically related group of *Mycobacterium* that cause tuberculosis
MVL	Mercury vapor lamp
NAAT	Nucleic acid amplification test
NK	Natural killer
NEAR	Nicking enzyme amplification reaction
NRA	Nitrate reductase assay
NTM	Non-TB mycobacteria
PCR	Polymerase chain reaction
PEPFAR	United States President's Emergency Plan for AIDS Relief
PET	Positron emission tomography
POCT	Point-of-care testing
PPD	Purified protein derivative
PTB	Pulmonary tuberculosis
RAM	Ramification-extension amplification
RCA	Rolling circle amplification
RD1	Region of difference
RNA	Ribonucleic acid
RPA	Recombinase polymerase amplification
SDA	Strand displacement amplification
SMART	Simple method for amplifying RNA targets
SmartAmp	Smart amplification process
TB	Tuberculosis
TBLB	Transbronchial lung biopsy
TLA	Thin layer agar
TMA	Transcription mediated amplification
TNFα	Tumor necrosis factor alpha
TPE	Tuberculous pleural effusion
TST	Tuberculin skin test
USAID	United States Agency for International Development

VOC	Volatile organic compounds
WHO	World Health Organization
XDR-TB	Extensively drug-resistant tuberculosis
ZN	Ziehl–Neelsen staining

Chapter 1
Introduction

Abstract The purpose of diagnostic guidelines for tuberculosis (TB) is to describe an acceptable level for all public and private professionals; they should try to achieve the proper diagnosis of patients who have, or suspected of having, or are at increased risk of developing TB. The basic principles of care for persons with, or suspected of having, TB are the same worldwide: a diagnosis should be established promptly and accurately; should use standardized treatment regimens proven, along with appropriate treatment support and supervision; response to treatment should be monitored; and the essential public health responsibilities must be carried out. The contribution of microbiology laboratory for diagnosis and management of TB is: (1) Collection of specimens for demonstration of tubercle bacilli; (2) Transport of specimens to the laboratory; (3) Digestion and decontamination of specimens; (4) Staining and microscopic examination; (5) Identification of mycobacteria directly from clinical specimens (nucleic acid amplification techniques (NAAT); (6) Cultivation of mycobacteria; (7) Identification of mycobacteria from culture; and (8) Drug susceptibility testing (DST). To identify *Mycobacterium tuberculosis* (MTB) in people without the disease, the methods used are the tuberculin skin test (TST) and the Interferon-gamma release assay (IGRA).

Keywords DST • IGRA • MTB • NAAT • TB • TST

Diagnostic guidelines are designed to provide a frame of reference to understand the diagnosis of infection/disease due to tuberculosis (TB), and present a classification outline that may simplify the management of individuals undergoing diagnostic testing.

The specific objectives of this volume are the following:

1. To define diagnostic strategies in high and low risk patient populations, based on current knowledge of the epidemiology of TB and information on new technologies.

This edition was prepared as a practical guide for microbiologists, physicians, health organizations, and educational institutions involved in the care of patients with TB. References have been included to guide the reader to texts and journal articles to obtain further detailed information on each subject.

© Springer International Publishing Switzerland 2015
G. García-Elorriaga, G. del Rey-Pineda, *Practical and Laboratory Diagnosis of Tuberculosis*, SpringerBriefs in Microbiology, DOI 10.1007/978-3-319-20478-9_1

2. Promote a precise diagnosis and hence, the administration of curative medical therapy.

The diagnosis of TB should hinge on a series of ancillary methods and confirmatory microbiological techniques. Ancillary diagnostic methods are nonspecific and include clinical manifestations, imaging findings (highly sensitive), histopathology (more specific than the other approaches), and the tuberculin skin test (TST) (a small contribution in terms of diagnosis).

When facing a possible case of disease due to TB, it is presumed that the infection by *M. tuberculosis* (MTB) has progressed enough to where clinical manifestations have developed, forcing the host/patient to seek medical help. Therefore, it is necessary to be familiar with the clinical picture of TB and with the relevant diagnostic techniques.

TB lacks specific clinical manifestations that can easily differentiate it from other respiratory diseases. In most cases, clinical manifestations in the TB setting are insidious and not particularly alarming; as a result, several months may go by before the diagnosis is established.

MTB may spread to any part of the body since it first enters the host. In summary, TB may affect any organ or tissue.

Aside from pulmonary involvement, the most common extrapulmonary sites of the disease include (in decreasing order): pleural, lymphatic, urogenital, osteoarticular, and meningeal, although as previously mentioned, any organ or tissue may be compromised. In immunocompetent patients, the frequency of extrapulmonary TB (EPTB) is no greater than 15–20 %, but this number increases in immune deficiency states as in the case of AIDS patients in whom extrapulmonary disease accounts for 50–60 % of all TB cases.

The physical examination of patients with TB disease also lacks specificity and frequently contributes scarce data to the diagnosis. In many cases, the patient appears healthy.

Also, laboratory testing does not provide characteristic information although tests should always be obtained for diagnostic purposes and in some cases, for patient follow-up during treatment.

The chest X-ray is a very sensitive technique in the diagnosis of PTB in immunocompetent individuals, although it is nonspecific since TB is not associated to pathognomonic radiological findings, regardless of how suggestive the images are. Although there are radiographic images that can greatly suggest the possibility of TB, these findings are only inferential support to the diagnosis and require the performance of confirmatory microbiological evaluations. The role of the chest X-ray in the TB diagnostic algorithm depends on the available resources and on the disease's prevalence in a specific population. In wealthy countries, radiological evaluation and microscopy are both recommended in all cases of suspected TB.

The value of the TST is very limited in the diagnosis of TB. But in children, especially if under the age of 5 and in whom the prevalence of MTB infection is very low, a positive test points to a very recent infection or real disease. In certain cases of a negative bacilloscopy and if EPTB hematogenous dissemination is suspected, biopsies may be required. The diagnosis is then based on the identification of caseous granulomas that may nevertheless be due to other disease processes, particularly as a result of other environmental mycobacteria or certain fungi.

Sampling and management of fluids and tissues affect the sensitivity of the different microbiological techniques. If possible, samples should be obtained before initiating chemotherapy and in open or well-ventilated areas.

In spite of the breakthroughs over the past 30 years in terms of microbiological TB diagnostic techniques, only a small segment of the population worldwide has benefitted. In low- and middle-income countries, the main diagnostic method is bacilloscopy with Ziehl–Neelsen's technique, due to its simplicity, rapidity, reproducibility, low cost, and efficacy in the detection of infectious cases.

The other basic TB diagnostic technique is culture, the only available method that can establish a definitive diagnosis and useful in patient follow-up and to ensure a definitive cure. Moreover, culture has greater sensitivity than bacilloscopy.

In terms of identification of the different mycobacterial strains, biochemical techniques such as chromatography and genetic probes are available. However, these biochemical tests are only routinely used in low- or middle-income countries due to their lower costs, although they may be complicated and slow to perform and they lack reproducibility.

In low- or middle-income countries, drug sensitivity must be tested following Canetti's proportions method and on Lowenstein–Jensen medium. Readout is conducted after 4–5 weeks and the laboratory must inform the physician on the amount of growth in media supplemented with anti-TB drugs compared to media with no drug.

The discovery of the structure of DNA, the elucidation of genetic molecular mechanisms, and the development of the polymerase chain reaction (PCR) have yielded powerful methods for the diagnosis of infectious disease. In spite of these advances, the gold standard for microbial identification is still culture and subsequent phenotypic differentiation of the causal pathogen.

TB diagnostic tests based on nucleic acid amplification (NAAT) are based on the amplification of specific short DNA or RNA sequences of the MTB complex with PCR; the amplified products are then detected by agarose/acrylamide gel electrophoresis or with various hybridization methods.

NAAT ensures a fast, sensitive, and specific diagnosis of infectious diseases. The next generation of diagnostic devices will focus on the genetic determinants of these disease at the point-of-care, so physicians can provide a quick and dependable diagnosis and hence, a more effective treatment.

Undoubtedly, molecular technology has yielded very powerful diagnostic tools. The diagnostics industry has recognized the many benefits of NAAT and is steadily investing in molecular diagnostic methods.

NAAT instrumentation has been miniaturized, fostering the equipment's portability. NAAT is particularly powerful in the prompt identification of pathogens whose prodromal presentation is similar but whose treatment strategies are entirely different.

In adults suspected of harboring TB, with or without associated HIV infection, it has been found that the MTB/RIF kit is sensitive and specific for TB detection. The Xpert MTB/RIF (GeneXpert) method is also valuable as a diagnostic tool in addition to bacilloscopy (sensitivity 67 %).

Using the GeneXpert kit for the diagnosis of multidrug-resistant TB (MDR TB) would cost less (US$70 to 90 million per year) than using a combination of conventional diagnostic methods (US$123 to 191 million per year). Conventional diagnostic methods include bacilloscopy, chest X-ray, culture, and drug susceptibility testing (DST) based on culture results and following the algorithms recommended by the WHO.

Specific tests are a strategy for TB control used to identify, evaluate, and treat individuals at high risk for latent TB infection (LTBI) or at high risk of developing the disease once infected with MTB. The identification of individuals with latent infection is pivotal for the control and elimination of TB, since treatment of LTBI may prevent infected individuals from developing the disease and also contribute to the arrest of TB propagation.

Other services (i.e., correctional institutions) must consult their local health departments before initiating a screening program in order to ensure the availability of resources for the evaluation and treatment of individuals testing positive for latent infection or TB disease. There are currently two methods for TB detection in the United States: (1) Mantoux TST and (2) Interferon-gamma release assay (IGRA).

A negative reaction to either test does not preclude LTBI or TB disease. Medical management or public health decisions must take into account clinical, epidemiological, historic, and other information when IGRA or TST is positive; the decisions should not be solely based on the TST or IGRA results.

TST is used to determine whether an individual is infected with MTB. In this test, a substance known as purified protein derivative (PPD) derived from tuberculin is injected under the skin. After 2–8 weeks since the initial TB infection, the immune system reacts to the PPD and the infection can be detected with the TST. TST involves the intradermal injection of 0.1 mL PPD containing 5 tuberculin units, on the volar surface of the arm.

The TST must be read 48–72 h after the injection, by a health worker trained in TST result interpretation.

The TST is read by palpation of the injection site, searching for an area of induration (firm swelling). The indurated area's diameter should be measured on the forearm.

The interpretation of TST reactions hinges on the size (in millimeters) of the induration and the risk of acquiring the infection or of progression to TB disease if the individual is infected. Individuals that have been previously vaccinated with BCG may develop a positive reaction to TST even if they have no TB infection.

Some individuals have a negative TST reaction in spite of a present MTB infection. False negative TST reactions may occur if infection developed within 8 weeks of the skin test.

The booster effect is mostly present in previously infected elderly individuals whose ability to react to tuberculin has faded over time. The two-step test refers to a strategy used to decrease the probability of misinterpreting a booster reaction as a recent infection.

IGRA detects MTB infection by measuring the immune response to TB proteins in whole blood. IGRA is unable to differentiate between a latent infection and active disease.

IGRA may be used for surveillance purposes or to identify individuals that could probably benefit from treatment, including those at higher risk of TB disease progression if infected with MTB. IGRA may be used instead of (but not in addition to) TST in all situations in which the Centers for Disease Control and Prevention (CDC) recommend it as a TB infection diagnostic aid.

Chapter 2
Clinical Diagnosis

Abstract A complete medical evaluation for tuberculosis (TB) includes the following: (1) Medical history, (2) Physical examination, (3) Tests for TB Infection, (4) Chest radiograph, and (5) Diagnostic microbiology. Tuberculin skin testing (TST) is the most common method used to screen for latent *M. tuberculosis* infection (LTBI). In 2001, an interferon release assay (QuantiFERON-TB test) was approved by the Food and Drug Administration (FDA). Active tuberculosis (ATB) is considered as a possible diagnosis when findings in a chest radiograph of a patient being evaluated for respiratory symptoms are abnormal, as occurs in most patients with pulmonary tuberculosis (PTB). The radiographs may show the characteristic finding of infiltrates with cavitation in the upper and middle lobes of the lungs, including clinical suspicion and response to treatment. Traditionally, the first laboratory test used to detect ATB in a patient with an abnormal chest radiograph is examination of a sputum smear in search for acid-fast bacilli (AFB). Definitive diagnosis of TB requires the identification of *M. tuberculosis* (MTB) in the culture of a diagnostic specimen. The most frequent sample obtained from a patient with a persistent and productive cough is sputum. Sputum is obtained by bronchoscopy and bronchial washings or bronchoalveolar lavage. Newer and faster MTB diagnostic techniques include nucleic acid amplification (NAA) tests. With molecular biology methods, DNA and RNA are amplified thus facilitating rapid detection of microorganisms; these tests have been approved by the FDA. Biopsy and/or surgery are required to procure tissue samples for diagnosis of extrapulmonary TB (EPTB).

Keywords Chest radiograph • Diagnostic microbiology • Histopathology • Medical history • Physical examination • Tests for TB Infection

© Springer International Publishing Switzerland 2015   7
G. García-Elorriaga, G. del Rey-Pineda, *Practical and Laboratory Diagnosis of Tuberculosis*, SpringerBriefs in Microbiology,
DOI 10.1007/978-3-319-20478-9_2

2.1   A Complete Medical Evaluation for Tuberculosis (TB) Includes the Following Five Components

2.1.1   Medical History

When conducting a medical history, the clinician should ask if any symptoms of TB are or have been present; if so, for how long, and whether there has been any known exposure to individuals with infectious TB. Equally important is whether or not the individual has been diagnosed in the past with latent TB infection (LTBI) or classic TB. Clinicians should determine if the patient has any underlying medical conditions, especially human immunodeficiency virus (HIV) infection or diabetes, that increase the risk of progression to active TB in those latently infected with *M. tuberculosis* (MTB) [6].

TB most commonly affects the lungs and is referred to as pulmonary TB (PTB). PTB often leads to general signs and symptoms, including cough (especially if lasting for 3 weeks or longer) with or without sputum production, coughing up blood (hemoptysis), chest pain, loss of appetite, unexplained weight loss, night sweats, fever, and fatigue.

Extrapulmonary TB (EPTB) may cause symptoms related to the compromised organ or system. For example, TB of the spine may cause back pain, TB of the kidney may be manifested as blood in the urine, or meningeal TB may lead to headache or confusion. Both PTB and EPTB symptoms can be caused by other diseases but should always prompt the clinician to consider a diagnosis of TB.

2.1.2   Physical Examination

The physical examination is an essential part in the evaluation of any patient. It cannot be used to confirm or rule out TB, but it can provide valuable information on the patient's overall condition, point to a specific diagnostic method and reveal other factors that may affect the treatment of TB, if so diagnosed. Physical findings are usually absent in mild or moderate disease. Dullness with decreased fremitus may indicate pleural thickening or effusion. In many instances, the patient appears to be healthy. Nevertheless, a systematic examination is always required in the search of possible diagnostic clues such as:

Crackling rales in the infraclavicular space or in the interscapulo-vertebral area due to exudative and cavitary lesions.

Uni- or bilateral bronchial rales (rhonchi, subcrepitations) in case of bronchogenic disease dissemination.

In case of pleural involvement, there is dullness on percussion and absence or decrease of vesicular murmur.

2.1.3   Test for TB Infection

Selection of the most suitable tests for detection of MTB infection should be based on the reasons and the context for testing, test availability, and overall cost-effectiveness. Currently, there are two available methods for the detection of MTB infection: Mantoux tuberculin skin test (TST) and Interferon-gamma (IFNγ) release assays (IGRAs), such as the enzyme-linked immunosorbent assay (QuantiFERON-TB Gold In-Tube (QFT-GIT), Cellestis Limited, Carnegie, Victoria, Australia) and the enzyme-linked immunospot assay (T-SPOT®.TB, Oxford Immunotec, Oxford, UK) [5, 8].

TST is the standard method of determining whether a person is infected with MTB. Reliable administration and reading of the TST requires standardization of procedures, training, supervision, and practice in Chap. 5.

For the first time, an alternative to the TST has emerged in the form of a new type of in vitro T-cell-based assay: IGRA. IGRAs are based on the principle that the T-cells of individuals sensitized with TB antigens produce IFNγ when they reencounter mycobacterial antigens in Chap. 5 [2, 3].

2.1.4   Chest X-Ray

All patients with persistent cough of more than 3 weeks duration should have a chest radiograph to rule out, among other diseases, PTB. Since PTB is the most common form of the disease, a chest X-ray (CXR) is useful in its diagnosis of TB. Chest abnormalities can suggest pulmonary TB. In some instances, a computerized tomography (CT) scan may provide additional information. A CT scan provides more detailed images that cannot be easily seen on a standard chest radiograph; however, CT scans can be substantially more expensive. In PTB, radiographic abnormalities are often seen in the apical and posterior segments of the upper lobe or in the superior segments of the lower lobe. However, lesions may appear anywhere in the lungs and may differ in size, shape, density, and cavitation, especially in HIV-infected and other immunosuppressed hosts. Often, the only vestige of a primary infection is a positive TST and the Ranke complex (Fig. 2.1).

The clinical and radiological suspicion of PTB is sufficient to initiate treatment without awaiting the culture result, but sputum should be obtained before the administration of therapy.

Mixed nodular and fibrotic lesions may contain slowly multiplying tubercle bacilli with the potential for progression to full-blown TB. Individuals with radiographic lesions suggesting "old" TB and with a positive TST reaction or positive IGRA should be considered high-priority candidates for treatment of LTBI, but only after active TB (ATB) is excluded in three specimens tested for acid-fast bacilli (AFB) by smear and culture because "old" TB cannot be differentiated from ATB on the basis of radiographic findings alone.

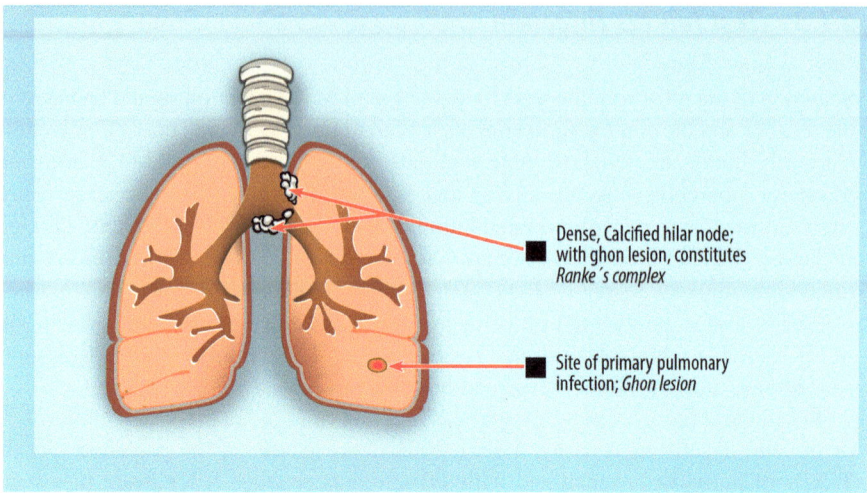

Dense, Calcified hilar node; with ghon lesion, constitutes *Ranke´s complex*

Site of primary pulmonary infection; *Ghon lesion*

Fig. 2.1 Radiographic residuals of primary infection

In HIV-infected individuals, PTB may present with atypical findings or with no obvious lesions on the CXR. The radiographic appearance of PTB in individuals infected with HIV might be typical; however, cavitary disease is less common among such patients. Abnormalities on CXR radiographs may be suggestive of, but are never diagnostic of TB. CXR may be used to exclude PTB in an HIV-negative patient who has a positive TST reaction or IGRA and no symptoms or signs of TB. For practical purposes, a normal CXR excludes TB.

The following CXR shadows are strongly suggestive of TB:

Upper lung patchy or nodular shadows (on one or both sides).
Cavitation (particularly if there is more than one cavity). Calcified shadows may lead to diagnostic difficulties. Remember that pneumonia and lung tumors can occur in areas of previous healed and calcified TB. Some benign tumors may also be calcified.

Other shadows that may be due to TB are:

Oval or round solitary shadow (tuberculoma).
Hilar and mediastinal shadows due to enlarged lymph nodes (persisting primary complex).
Diffuse, small, nodular shadows (miliary TB).

The correct reading of a CXR requires significant experience. If you suspect TB based on the X-ray but the sputum is negative, administer a non-TB antibiotic (e.g., ampicillin, oxytetracycline) for 7–10 days and then obtain another CXR. Shadows due to acute pneumonia will have improved.

Table 2.1 Usefulness of chest radiography as a diagnostic test for TB

Radiographic finding	Sensitivity (%)	Specificity (%)
Any abnormalities consistent with TB (active or inactive)	98 (95–100)	75 (72–79)
Abnormalities suggestive of active TB	87 (79–95)	89 (87–92)
After screening positive symptoms (one study)	90 (81–96)	56 (54–58)

In individuals with a negative smear and PTB symptoms, an abnormal CXR may be very useful diagnostically. However, a diagnosis of TB can not be established only by radiography. Although the sensitivity of CXR is high, the specificity is low, as shown in Table 2.1.

2.1.5   Bacteriologic Examination of Clinical Specimens

Examination of clinical specimens (e.g., sputum, urine, or cerebrospinal fluid) is of critical diagnostic importance. The specimens should be examined and cultured in a laboratory that specializes in testing for MTB. The bacteriologic examination includes five stages: (1) Specimen collection, processing, and review; (2) AFB smear classification and results; (3) Direct detection of MTB in the clinical specimen with a nucleic acid amplification test (NAAT); (4) Specimen culture and identification; and (5) Drug susceptibility testing.

2.2   Other Tests

2.2.1   Adenosine Deaminase (ADA)

EPTB accounts for 10 % of all cases and pleural TB is the second most common manifestation, preceded only by lymphonodular TB. The diagnosis of the first condition is established by bacteriological means and identifying the bacillus in pleural fluid; unfortunately, staining when in search of MTB is usually negative and culture is positive in less than 25 % of cases. Furthermore, a pleural biopsy only shows granulomatous pleuritis in 80 % of patients with tuberculous pleural effusion (TPE) [4]; biopsy culture combined with histological examination, establishes the diagnosis in approximately 90 % of cases.

Measurement of ADA activity has proven to be sensitive (73 %) and specific (90 %) for pleural TB in special circumstances, such as in regions with a high prevalence of TB. The levels of ADA, an enzyme found in most cells, are increased in TPEs; this determination has acquired popularity as a diagnostic test in areas with a high incidence of TPE because it is noninvasive, the assay is not expensive, and it is readily accessible. The demonstration of elevated pleural fluid ADA levels is useful in establishing the diagnosis of tuberculous effusions. ADA is an enzyme involved in purine catabolism, catalyzing the conversion of adenosine to inosine. The colorimetric test is based on the quantification of ammonium yielded as a result of the enzymatic activity. The reported cutoff value for ADA varies from 47 to 60 U/L. Specificity is increased when the lymphocyte/neutrophil ratio in pleural fluid (>0.75) is weighed in association with an ADA concentration >50 U/L. Exudative lymphocytic pleural effusions are commonly encountered in clinical practice and are often a challenging diagnostic problem. The two most common causes are malignancy and tuberculous effusions, making the test less useful in countries with a low prevalence of TB.

With the declining prevalence of PTB, the positive predictive value of pleural fluid ADA has also decreased although its negative predictive value has actually increased. Therefore, the measurement of pleural fluid ADA levels can be used to rule out a tuberculous etiology of lymphocytic pleural effusions, regardless of the prevalence of TB.

This test is mostly used to confirm the diagnosis if disease is suspected rather than to rule it out.

2.2.2 Histopathology

The multiplication of tubercle bacilli at any site causes a specific type of inflammation, with formation of the characteristic granuloma (Fig. 2.2). Pathology analysis involves examining the tissue for suspected TB [1]. Histopathology entails the microscopic examination of a tissue sample and it is a diagnostic aid when bacteriological techniques cannot be applied. It is especially useful for EPTB.

Types of samples: (1) Aspiration of the lymph nodes; (2) Biopsy of the serous membranes; and (3) Tissue biopsy: (a) Without surgery; (b) During surgery; and (c) Postmortem.

Methods: (1) Cytological techniques; (2) Bacteriological and histological techniques on biopsied samples; (3) Bacteriological techniques; and (4) Histology techniques.

Practical point: on biopsy, at least two fragments should be collected; one is placed in saline solution and transferred to the mycobacteriology laboratory for culture, while the other undergoes fixation for histological evaluation.

Microscopic aspects: Organ involvement by TB leads to an inflammatory reaction in the affected site. Inflammation develops in three successive stages that can overlap—acute, subacute, and chronic—and that have different histological characteristics.

Practical point: among all types of lesion, only follicular lesions with necrotizing granulomas are specific enough to confirm the diagnosis of TB, as is detection of bacilli on histological samples after appropriate staining.

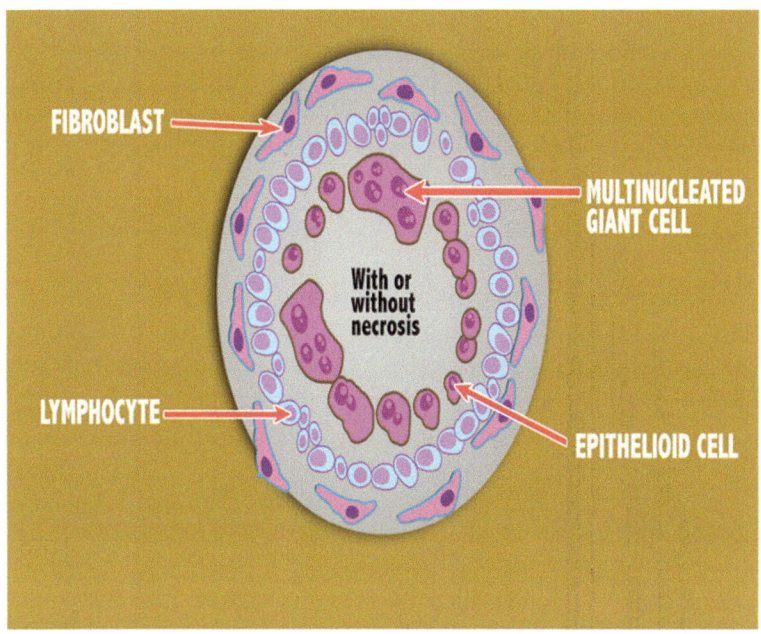

Fig. 2.2 Granuloma. Focal accumulation of macrophages (histiocytes), and/or modified (epithelioid) surrounded by a collar of lymphocytes, surrounded by a ring of fibroblasts

The histopathological examination of biopsies obtained by bronchoscopy reveals caseating granulomatous inflammation in all patients.

Renal and urinary TB diagnosis is established with certainty by identifying the Koch bacillus in special cultures and using histological examination of surgically obtained tissue. Although not in all cases, the usual pathological techniques tend to be very specific. Morphological changes are initially found in the renal cortex and granulomas subsequently develop in the medulla. The pathognomonic finding is a central area of caseating necrosis and eosinophilic nonstructured infiltrates with necrotic detritus surrounded by a row of epithelioid macrophages and a few giant cells: this characterizes the classical caseating granuloma.

The necrotic area consists of amorphous, pink, caseous material composed of the granuloma's necrotic elements as well as infectious organisms. The epithelioid macrophages are elongated with long, pale nuclei and pink cytoplasm while macrophages are grouped in what are known as giant cells. The typical giant cell in infectious granulomas is known as a Langhans giant cell, with the nuclei lined up along one edge of the cell. The necrotic area is surrounded by the inflammatory component of epithelioid cells, lymphocytes, and fibroblasts.

Although bacteriology remains the key test confirming the diagnosis of TB, histology does play an important role, particularly in the diagnostic confirmation of extrapulmonary variants. Combining histological techniques with bacteriology increases the rate of histological diagnoses. Bacteriological culture of tissue fragments (or albeit less useful, fluid) obtained concurrently with the samples for histological analysis, increases the rate of EPTB confirmed diagnoses [7].

In cases of tuberculous cervical lymphadenitis, characteristic epithelioid cell granulomas with central caseating necrosis are seen in all the cases. Analysis by light microscopy is still a useful screening method in the diagnosis of TB cervical lymphadenopathy.

2.3   Evaluation of Diagnostic Methods in EPTB

It is recommended to obtain a sample, if possible, directly from the site of concern; if necessary, this is accomplished by needle or fine-needle aspiration and the tissue/fluid should be sufficient for histology analysis, smear, and culture. Various imaging tests are recommended, depending on the compromised organ or system, in the diagnosis of suspected EPTB. However, a CXR should always be obtained to rule out a pulmonary component. Aside from the microbiological and histological study of the sample, a rapid diagnostic technique is recommended in cases where treatment should be promptly initiated, as in tuberculous meningitis or severe disseminated TB.

2.3.1   Diagnosis of Miliary TB

The diagnosis of miliary TB can be difficult since clinical manifestations are non-specific, the chest radiographs do not always reveal classical miliary changes, and patients may present with complications that may distract the clinician. Therefore, a high index of clinical suspicion and a systematic approach to diagnostic testing is pivotal in establishing the diagnosis of miliary TB. The following criteria are useful in the diagnosis of miliary TB: (1) Clinical presentation consistent with a diagnosis of tuberculosis; (2) Classical miliary pattern on chest radiograph; (3) Bilateral diffuse reticulonodular lung lesions on a background of miliary shadows demonstrable either on plain chest radiograph or high-resolution computed tomography (HRCT); and (4) Microbiological and/or histopathological evidence of TB. Tuberculin anergy is more common in miliary TB (20–70 %) than in pulmonary and EPTB; TST conversion may occur following successful treatment.

A miliary pattern on the CXR is often the first clue suggesting miliary TB. Several other imaging modalities such as ultrasonography, computed tomography (CT), magnetic resonance imaging (MRI), or positron emission tomography (PET) can further assess the extent of organ involvement and are also useful in evaluating response to treatment. A miliary pattern on the chest radiograph is the hallmark of miliary TB and it is evident in most patients. If caseous material, collagen, or both are present in the tubercles, they are visible in the CXR. A classical miliary pattern on the chest radiograph represents the summation of all densities in tubercles that are perfectly aligned, whereas those that are imperfectly aligned result in curvilinear densities and a reticulonodular pattern.

In comparison with the pre-CT era, HRCT and thin-section multidetector CT (MDCT) have facilitated the antemortem diagnosis of miliary TB. With the availability of these imaging modalities, cryptic miliary TB that previously could have only been diagnosed at autopsy can now be diagnosed in a timely manner. The HRCT reveals a composite of both sharply and poorly defined nodules, <2 mm and that are widely disseminated throughout the lungs in association with diffuse reticulation. Importantly, the HRCT may reveal a classical miliary pattern even when the chest radiograph is apparently normal and also facilitates the identification of additional findings such as intrathoracic lymphadenopathy, calcification, pleural, and pericardial lesions [9]. CT and MRI have been useful in identifying miliary lesions in extrapulmonary sites. Abdominal CT has been useful in identifying lesions in the liver, spleen, intestine, mesentery, peritoneum, adrenals, and lymph nodes and can also detect cold abscesses. Brain and spine MRI is very useful in the evaluation of patients with miliary TB, MTB, and spinal TB. MRI is particularly helpful in identifying and delineating the extent of tuberculomas and cold abscesses as well as monitoring the response to treatment.

Although not all patients with miliary TB have a productive cough, if present, sputum must be collected for smears and mycobacterial culture. Sputum smear microscopy using the Ziehl–Neelsen stain is useful in detecting AFB. Fiberoptic bronchoscopy, bronchoalveolar lavage (BAL), bronchoscopic aspirate, brushings, washings, and transbronchial lung biopsy (TBLB) are also useful in confirming the diagnosis of miliary TB. The cumulative diagnostic yield for various bronchoscopically obtained specimens analyzed by smear and culture has been found to be 50 %. Depending on the extent of organ system involvement, appropriate tissue and body fluid samples must be obtained to confirm the diagnosis from a histopathological and microbiological perspective.

The World Health Organization (WHO) policy statement on the use of serodiagnostic tests strongly recommends that currently available commercial tests should not be used for the diagnosis of active pulmonary and EPTB including miliary TB. ADA and interferon-gamma levels in ascitic and/or pleural fluid can be helpful in the diagnosis of miliary TB. In patients with suspected miliary TB, automated molecular tests for MTB detection and drug-resistance testing may be used for early confirmation of the diagnosis, if these techniques are available. Based on current evidence and expert opinions, molecular assays to detect gene mutations signaling drug resistance have been endorsed by the WHO as being most suited for a rapid diagnosis.

Miliary TB is associated with typical interstitial lung disease abnormalities in pulmonary function tests. Diffusion impairment is the most common abnormality and may sometimes be severe. Abnormal cardiopulmonary exercise performance has been described in patients with miliary TB. Salient abnormalities include decreased maximum oxygen consumption, maximal work rate, anaerobic threshold, peak minute ventilation, breathing reserve, and a low maximal heart rate.

2.3.2   Diagnosis of Pleural TB

The determination of ADA in pleural fluid has a sensitivity of 73 % and a specificity of 90 %. In high prevalence countries, the probability of diagnosing pleural TB after determining ADA is 99 %.

Screening for interferon in pleural fluid has been compared in culture vs. histology, yielding a sensitivity and specificity of 89 % (CI 95 % 87–91) and 97 % (CI 95 % 96–98), respectively.

The technique of nucleic acid amplification (NAAT) has been evaluated in patients with pleural TB. Marketed tests show a combined sensitivity of 62 % and a specificity of 98 % but these results tend to be heterogeneous.

2.3.3   Diagnosis of Meningeal TB

NAAT tests have been evaluated in the diagnosis of MTB. The overall results revealed a relatively low sensitivity (71 %) but with a good specificity (95 %) and all with a significant variability [10].

ADA was evaluated by comparing the results obtained by culture, clinical skills, or histology. The results are very heterogeneous and sensitivity varies between 36 and 100 %, while the specificity ranged between 63 and 99 %.

The diagnostic performance of IGRA techniques is inferior in tuberculous meningitis.

2.3.4   Diagnosis of Pericardial TB

ADA validity for diagnosing pericardial TB has been evaluated. A cutoff value (47 U/L) was established and revealed a sensitivity of 88 % and a specificity of 83 %. The gold standard was the identification of the bacillus or the clinical outcome.

2.3.5   Diagnosis of Lymph Node TB

Diagnostic NAAT techniques (commercial or not) were evaluated in patients with lymph node TB. Its sensitivity ranged between 2 and 100 % and its specificity was 28–100 %. The performance of commercial tests was superior.

2.3.6   Diagnosis of Abdominal TB

Determining ADA in the ascitic fluid is also used in the diagnosis of abdominal TB. It has been used in patients with peritoneal TB for microbiological (smear and culture) and/or histological diagnosis. Its sensitivity was 100 % and its specificity was 97 %. Although ADA determination in ascites may prevent aggressive exploration and laparoscopy, false positive results may be due to serious diseases, such as different types of malignancies.

2.3.7   Diagnosis of Resistance to Anti-TB Drugs

Automated methods in liquid culture media (MGIT960, MB/BacT ALERT 3D, and Versa TREK) are currently the most commonly used due to their speed and reliability. These methods correctly detect multidrug resistance to isoniazid and rifampicin although there is some variability between laboratories in terms of the detection of resistance to other first-line drugs such as pyrazinamide, ethambutol, or streptomycin.

More recently, methods have been developed for rapid detection of resistance in clinical samples. These methods include the detection of bacteriophages with luciferase reporter phage (LRP) methods and the mycobacteriophage-based assay (MBA); both are extremely useful in the clinic.

First-line drug sensitivity in the initial isolates from all patients with TB should be tested. Susceptibility testing of second-line drugs should be conducted if microbiological resistance is established or if there is suspected clinical resistance to first-line drugs, as in case of failure in the initial response or after relapse once treatment is completed. Sensitivity studies should be carried out in laboratories with accredited quality controls.

References

1. Ait-Khaled N, Enarson DA. Tuberculosis a manual for medical students. WHO/CDS/ TB/99.272. Geneva: WHO; 2003.
2. García-Elorriaga G, Martínez-Velázquez M, Gaona-Flores V, del Rey-Pineda G, González-Bonilla C. Interferon γ in patients with HIV/AIDS and suspicion or latent tuberculosis infection. Asian Pac J Trop Med. 2013;6:135–8.
3. Grant J, Jastrzebski J, Johnston J, Stefanovic A, Jastrabesky J, Elwood K, et al. Interferon-gamma release assays are a better tuberculosis screening test for hemodialysis patients: a study and review of the literature. Can J Infect Dis Med Microbiol. 2012;23:114–6.
4. Kelam MA, Ganie FA, Shah BA, Ganie SA, Wani ML, Wani NU, et al. The diagnostic efficacy of adenosine deaminase in tubercular effusion. Oman Med J. 2013;28:417–21. doi:10.5001/ omj.2013.118.

 5. Lalvani A, Pareek M. Interferon gamma release assays: principles and practice. Enferm Infecc Microbiol Clin. 2010;28:245–52. doi:10.1016/j.eimc.2009.05.012.
 6. National Center for HIV/AIDS, Viral Hepatitis, STD, and TB Prevention. Core curriculum on tuberculosis: what the clinician should know. 6th ed. Atlanta: Division of Tuberculosis Elimination, Department of Health & Human Services, CDC; 2013.
 7. Ozkaya S, Bilgin S, Findik S, Kök HC, Yuksel C, Atıcı AG. Endobronchial tuberculosis: histopathological subsets and microbiological results. Multidiscip Respir Med. 2012;7:34. http://www.mrmjournal.com/content/7/1/34. Accessed 19 Feb 2014.
 8. Pai M, Riley LW, Colford Jr JM. Interferon-γ assays in the immunodiagnosis of tuberculosis: a systematic review. Lancet Infect Dis. 2004;4:761–76.
 9. Sharma SK, Mohan A, Sharma A. Challenges in the diagnosis & treatment of miliary tuberculosis. Indian J Med Res. 2012;135:703–30.
10. TB CARE I. International standards for tuberculosis care. 3rd ed. The Hague: TB CARE I; 2014.

Chapter 3
Bacteriological Diagnosis

Abstract Sample collection and handling influences the sensitivity of the different microbiological techniques employed. Whenever possible, samples should be collected before beginning chemotherapy, in open areas or in well-ventilated rooms. Despite the advances made in the past 30 years in the microbiological techniques used for diagnosing TB, only a small portion of the global world population can benefit from them. The main diagnostic method in low- or middle-income countries is smear microscopy with the Ziehl–Neelsen technique, due to its simplicity, rapidity, reproducibility, low cost, and effectiveness in detecting infectious patients. The other basic diagnostic technique of TB is culture, the only available method that can establish a definitive diagnosis and that can be used for close patient follow-up and to confirm final cure. Culture, moreover, is more sensitive than smear microscopy. Biochemical techniques, chromatography, and genetic probes can be used to identify different mycobacterial species. However, only biochemical testing is routinely used in low- or middle-income countries due to its lower cost; these methods, however, may be complicated to perform; they yield results slowly and lack reproducibility. In low- or middle-income countries, drug sensitivity should be determined following Canetti's method of proportions in Löwenstein–Jensen medium. Reading should be obtained after 4–5 weeks, and the laboratory must inform the clinician on the sample's level of growth in the medium with added antituberculous drugs and in comparison to the medium without medication.

Keywords Acid-fast bacilli • Drug susceptibility testing • Löwenstein–Jensen culture • MB/BacT system • MGIT • MODS

3.1 Sampling Methods

The purpose of microbiological laboratory techniques is to isolate and identify causative organisms as well as to test antimicrobial susceptibility. The contribution of microbiology to the diagnosis of TB depends on the quality of the collected samples and the techniques employed. This section reviews the basic protocols for collecting, transporting, and processing samples, followed by a review of the different techniques.

© Springer International Publishing Switzerland 2015   19
G. García-Elorriaga, G. del Rey-Pineda, *Practical and Laboratory Diagnosis of Tuberculosis*, SpringerBriefs in Microbiology,
DOI 10.1007/978-3-319-20478-9_3

3.1.1 Importance of Sample Collection and Processing in Pulmonary TB

A series of norms have been established for the collection, storage, and shipment of samples. These guidelines must be followed consistently, since any deviation may compromise the sensitivity of the different microbiological techniques [1].

1. Whenever possible, samples should be obtained before initiating chemotherapy.
2. Samples should be obtained in open areas or well-ventilated rooms and away from other people.
3. Sputum and urine samples are to be shipped in sterile and properly identified, wide-lipped glass or plastic containers with airtight screw-on covers.
4. By international standards, in all patients (adults or children capable of producing sputum) with suspected pulmonary tuberculosis (PTB), at least three consecutive sputum specimens are required and should be collected in 8–24-h intervals; at least one sample should be an early morning specimen.
5. The time between sampling and analysis should be minimal but if it exceeds 1 h, the sample should be stored in the refrigerator at about 4 °C.
6. There are different sputum collection techniques: spontaneous expectoration, gastric aspirate (children), induced sputum production, and bronchoscopy.
7. All biopsy samples should be transferred to the microbiology laboratory without fixation and with only a few added drops of distilled water to prevent dehydration. Formalin should be avoided, although it is appropriate for samples evaluated by the pathology laboratory.
8. In HIV-infected patients, in whom disseminated TB is much more frequent, collection of all possible samples (including sputum, urine, cerebrospinal fluid, organ biopsy specimens) to confirm the diagnosis should be conducted. In patients with severe immune deficiency and fever of unknown origin, three blood cultures to search for *Mycobacterium tuberculosis* (MTB) may be adequate.

3.1.2 Specimen Collection Methods in Extrapulmonary TB

TB can occur in almost any anatomical site; thus, a variety of clinical specimens other than sputum (e.g., urine, cerebrospinal fluid, pleural fluid, pus, or biopsy specimens) must be obtained for examination when extrapulmonary TB (EPTB) is suspected. Procedures for the expeditious and recommended handling of the specimen must be in place or assured before the specialist performs an invasive procedure to obtain the specimen.

3.1.3 Acid-fast Bacilli Smear Classification and Results

Detection of acid-fast bacilli (AFB) in stained and acid-washed smears examined microscopically in a clinical specimen may provide the initial bacteriological evidence for the presence of mycobacteria. Smear microscopy is the quickest and easiest procedure that can be performed.

There are two commonly used procedures for acid-fast staining: (1) carbolfuchsin methods, including the Ziehl–Neelsen (ZN) and Kinyoun methods (direct microscopy) and (2) the fluorochrome procedure using auramine-O or auramine–rhodamine dyes (fluorescent microscopy).

Studies have shown that there must be 5000–10,000 bacilli/mL of specimen to detect bacteria in stained smears. In contrast, only 10–100 bacilli are needed for a positive culture. Many TB patients have negative AFB smears and a subsequent positive culture. Negative smears do *not* exclude TB.

If acid-fast bacilli are seen in the smear, they should be counted. A system is in place to report the number of acid-fast bacilli observed at certain magnifications.

3.1.4 Evolution of the Microbiological Techniques Used to Diagnose Tuberculosis

The evolution of microbiological diagnostic techniques since Robert Koch first used smear microscopy in 1882 has gone through four well-differentiated phases in which significant progress has been haphazard.

1. It was characterized by few advances: laboratories used conventional technology with limitations due to smear microscopy's low sensitivity, the excessively prolonged duration of culture, microorganism identification and susceptibility testing.
2. The introduction of new culture technology that to date has yet to be improved: the so-called radiometric growth detection systems. The main limitation of this new technology was the need to work with radioactive isotopes, a major obstacle for many laboratories that had no license to store and work with such materials.
3. The advent of HIV and the accelerated development of new technologies such as the rapid non-radiometric culture techniques, the standardization of effective systems to isolate mycobacteria from blood (blood cultures), and the development of rapid identification techniques (e.g., genetic probes, chromatography) as alternatives to traditional biochemical methodology.
4. The development of new genetic amplification techniques for the rapid diagnosis of TB.

Still, these important advances in microbiological diagnosis are used almost exclusively in the richest countries, since their cost and complexity render them unfeasible in poorer countries with paradoxically, the greatest TB burden.

3.1.5   Conventional Microbiological Techniques in the Diagnosis of Tuberculosis

Conventional microbiological techniques when diagnosing TB are the only routinely recommended methods in low- and middle-income countries. Only in exceptional situations (discussed later) are other techniques justified.

The conventional microbiological diagnosis of TB is based on four successive stages: (1) Sample staining for direct visualization under the microscope (smear microscopy); (2) Solid medium culture; (3) Identification of the microorganism with biochemical techniques; and (4) Drug susceptibility testing (DST) [7, 8].

3.1.6   Smear Microscopy

Microscopy is the simplest and quickest currently available procedure to detect AFB in clinical specimens, using the classic ZN staining method or one of its variants [3]. However, AFB detection is limited with this method since it requires at least 5 to 10×10^3 bacilli/mL of sputum.

MTB is a Gram-positive or frequently colorless bacterium, so it is often not visualized in routinely processed samples.

The detection of AFB in stained preparations examined under the microscope is the first evidence of mycobacteria in a clinical sample. The acid-fast characteristics of the microorganism are attributable to the high lipid content of the bacterial cell wall.

For over a century, sputum smear examination for AFB detection by ZN method and Löwenstein–Jensen culture has been the main TB diagnostic tool. Sputum smear examination is reliable, reproducible, and cost-effective for TB diagnosis and to monitor patient response to anti-TB treatment.

The results of smear microscopy can be influenced by the type of specimen, the thickness of the smear, the degree of discoloration, the type of counter stain used as well as the examiner's training and experience.

The sensitivity of smear microscopy is relatively limited. This implies that a negative result does not exclude the disease, since many false negative results are possible. Several approaches that have been implemented to increase its sensitivity are classical techniques (ZN and Kinyoun) and fluorescence smears. The sensitivity of conventional methods varies between 32 and 94 % while that of fluorescence is between 52 and 97 %.

The advantages of smear microscopy:

1. Simplicity and reproducibility in any setting.
2. Results can be reported within hours of obtaining the sample and they are reliable epidemiological indicators required for the evaluation of TB control.
3. It is inexpensive.
4. It is highly specific.
5. It detects MTB transmitters.

In developing countries, smear microscopy is likely to remain for the foreseeable future, the only cost-effective tool to diagnose patients with infectious TB and to monitor treatment progress.

The acid-fast staining characteristics observed with smear microscopy are common to all species of the genus *Mycobacterium*, as well as to some fungal species. As a result, all environmental mycobacteria look the same under the microscope.

Sometimes, the culture becomes negative before smear microscopy because previously administered therapy makes the bacilli nonviable. However, mycobacteria continue to be eliminated by the host and still exhibit acid-fast staining characteristics.

3.1.7 Auramine–Rhodamine Fluorescent Staining

This is the stain of choice when the number of samples to be examined is large. It is based on the affinity of the mycobacterial cell wall's mycolic acids for fluorescent stains or the auramine–rhodamine fluorochromes.

In fluorochrome staining, the first stain may be auramine, rhodamine or a combination of both (auramine–rhodamine). Auramine–rhodamine yields the best results [9]. Once fixed to the bacteria, these appear yellow or bright orange in a dark background.

One of the main advantages of fluorochromes is that they allow AFB detection in less time than with the conventional ZN technique, since the bacilli can be observed at low magnification (×20, ×25, or ×40) and morphology is subsequently confirmed by immersion microscopy (×100). Another advantage of this technique is that it allows smear restaining for Ziehl–Neelsen to determine and confirm the microorganism's morphology.

When reading the smear, the microscopist should provide the clinician with an approximate estimate of the number of detected AFB. Table 3.1 shows the frequently used guideline to quantify the microorganisms seen in the bacilloscopy.

Table 3.1 Acid-fast bacilli quantification scale according to stain

Carbolfuchsin (×1000)	Fluorochrome (×250)	Reported numbers
No AFB/300 fields	No AFB/30 fields	No AFB seen
1–9/100 fields	1–9/10 fields	Few (1+)
1–9/10 fields	1–9/field	Moderate (2+)
1–9/field	10–90/field	Many (3+)

3.1.8 Light-Emitting Diode Microscopy

The World Health Organization has advised that light-emitting diode (LED)-powered fluorescence microscopes replace mercury vapor lamp (MVL)-powered fluorescence microscopes for the detection of AFB stained with auramine. The performance of the Fluo-RAL module (RAL Diagnostics Company) has been evaluated and it appears to be a less dangerous and more durable source of light than the MVL for the microscopic detection of auramine-stained AFB [2].

LED is a low-cost method that offers the benefit of fluorescence microscopy without the associated operational requirement of a dark room and a special microscope; it also may be battery operated.

There are various methods to process a sputum sample in order to increase its sensitivity when detecting bacilli microscopically. Centrifugation in association with any chemical homogenization method increases sensitivity more than sedimentation.

The diagnostic performance of sputum serial samples has also been evaluated. Weighed analysis studies have determined that the gold standard is culture with a first analyzed sputum sensitivity of 53.8 %.

3.2 Culture Methods

Isolation of mycobacteria from clinical samples by culture is still the cornerstone on which the definitive diagnosis of TB and other mycobacterioses relies. At present, mycobacterial culture can be performed on conventional egg-based solid media such as Löwenstein–Jensen (LJ) and on those based on agar such as Middle brook 7H10 or 7H11, and in liquid media such as Middle brook 7H9 broth.

Although a combination of solid and liquid media is currently the gold standard for the primary isolation of mycobacteria, a few modern, rapid methods are also available. These include microscopic observation drug susceptibility (MODS), the BACTEC-TB460 radiometric system, and the Mycobacteria Growth Indicator Tube (MGIT).

3.2.1 LJ Culture

This is the gold standard for the diagnosis of TB and it also facilitates DST.

Mycobacterial culture is the only acceptable method available for patient follow-up and to confirm cure. For this reason, in countries with the sufficient economic resources, all clinical samples suspected of containing mycobacteria should be grown in the appropriate culture media.

The results of culture are largely dependent on sample decontamination and digestion because most clinical samples contain abundant commensal flora that grow faster than MTB.

Culture offers several advantages that define it as the gold standard in the diagnosis and follow-up of TB cases. These advantages can be summarized as follows:

1. Cultures are much more sensitive than smear microscopy and are able to detect as few as 10–100 bacteria/mL of sample.
2. Isolation in pure culture is necessary to correctly identify the isolated strains, since other mycobacteria appear identical to MTB by smear microscopy.
3. Culture provides definitive confirmation of negative conversion and healing of patients with treatment. In poor countries, where problems associated with treatment (e.g., suspected resistance) make culture-based follow-up necessary, the number of colonies obtained must be quantified; this parameter is pivotal for treatment monitoring and assessing possible treatment failure.

However, the logistical problems posed by culture methods limit its use, particularly in poorer countries. Its main inconveniences can be summarized as follows:

1. The most relevant limitation of conventional culture hinges on MTB's slow divisional capacity. This leads to a prolonged period between sample collection and the report of results when using conventional solid media—no less than 4–6 weeks and much longer in poor countries.
2. The cost of culture is far greater than that of smear microscopy; specific media are also needed as well as storage in an incubator. More specific training of personnel performing the cultures is also required.

In view of the above considerations and unlike smear microscopy, it is not possible to use culture at the most peripheral levels of health care. Thus, if TB is clinically suspected and smear microscopy proves positive in this setting, treatment should be initiated and the patient registered as a TB case.

The indication for culture is therefore dependent on the endemicity of the disease in the area and on the available health care infrastructure and resources. It can generally be concluded that in industrialized countries that for many years have successfully diagnosed cases by smear microscopy among mildly ill patients, and that have many health care centers and laboratories with no economic constraints, culture should be performed whenever a clinical sample is obtained from a patient with suspected TB.

An intermediate position is represented by middle-income countries, where patients with a positive smear microscopy are often appropriately managed and where economic constraints are not as severe as in poorer countries. Here, culture should be performed whenever therapy appears to be of no use (e.g., suspected failure, default, relapse).

Traditional culture has always been performed in solid medium, using coagulated egg (LJ) or agar (Middlebrook 7H10 and 7H11) as a base. These should be the only media routinely used in countries with low- or middle-income levels and preferably, LJ medium is recommended.

These solid media offer the advantages of simplifying culture, the counting of bacterial colonies (which is important in the follow-up of patients exhibiting a poor bacteriological response), detecting the growth of more than one mycobacterium in the sample, and they are also cost-effective. However, solid media are inconvenient in that bacterial growth is slow and results must be read manually, which could lead to errors.

3.2.2   Liquid Culture, DST

Implementation of liquid cultures has been recommended by the WHO since they yield higher rates of mycobacterial isolation and the time to detection is shorter than with classical solid media.

BACTEC-TB460 radiometric system measures the radioactive CO_2 released during decarboxylation of ^{14}C labeled substrates. If the inoculum contains live TB bacilli, they utilize the ^{14}C labeled substrate (palmitic acid) and release $^{14}CO_2$. The BACTEC instrument quantitatively measures the radioactivity; the daily increase in the growth index is directly proportional to the rate and amount of growth in the medium. By adding inhibitory substances to the medium, DST can also be performed.

3.2.3   MB/BacT System

This is a non-radiometric continuous monitoring system with a computerized database. It is based on the colorimetric detection of CO_2 [4].

When comparing the performance of the MB/BacT system with that of BACTEC-TB460, the mean time until detection of MTB by the BACTEC system was 11.6 days vs. 13.7 days by the MB/BacT system. The conclusion was that the MB/BacT with its computerized data management system is an acceptable alternative to BACTEC 460 despite some minor disadvantages such as increased contamination and a slightly longer time period until detection of growth.

3.2.4   The MGIT

This was initially introduced as a manual system and subsequently, as an automated BACTEC MGIT-960 system (Becton Dickinson, Sparks, MD, USA). This system uses tubes containing enriched Middlebrook 7H9 broth with an oxygen-sensitive fluorescence sensor embedded in silicone at the bottom of the tube and which upon consumption of the oxygen by the mycobacteria in the culture medium, fluoresces orange when probed with an UV light.

3.2.5   Non-commercial Culture Methods

The MODS

Assay uses a liquid medium that fosters faster growth of the TB bacillus and thereby aids in the early microscopic visualization of characteristic cord formations with an inverted light microscope. The comparison of growth in drug-containing and drug-free wells, susceptible or resistant strains allows anti-TB DST.

3.2.6   Newer Solid Cultures

The Nitrate Reductase Assay (NRA)

Is based on the principle of detection of nitrites due to the action of nitrate reductase (the Griess method). The accuracy of these methods has been evaluated in systematic reviews and meta-analyses confirming their high sensitivity and specificity in the detection of drug-resistant (DR)-TB strains.

The Thin Layer Agar (TLA) Culture

Uses a solid medium and is based on the detection of early mycobacterial growth according to colony morphology. The sample is inoculated on plates with Middlebrook 7H11 and Middlebrook 7H11 enriched with para-nitrobenzoic acid (PNB). The *M. tuberculosis* complex is expected to grow on the plate with Middlebrook 7H11, but not in Middlebrook 7H11 enriched with PNB, that inhibits its growth.

Colorimetric Redox Indicator (CRI)

This is a colorimetric detection method that, instead of looking for mycobacterial growth as colonies, detects the metabolic activity of the TB bacillus in a color reaction, using redox indicators such as resazurin and tetrazolium salts [3-(4,5-dimethylthiazol-2-yl)-2,5-diphenyl tetrazolium bromide].

3.3   Identification of Mycobacteria

Positive cultures for *M. tuberculosis* confirm the diagnosis of TB; however, in the absence of a positive culture, TB may also be diagnosed on the basis of clinical signs and symptoms alone. Cultures should be obtained in all diagnostic specimens, regardless of the AFB smear or nucleic acid amplification test NAAT results.

The mycobacteria comprising the MTB complex (MTBC) can easily be differentiated with a set of biochemical tests, since these microorganisms are niacin positive, reduce nitrates to nitrites, and possess pyrazinamidase (which allows the distinction between MTB and *M. bovis*) as well as a heat-sensitive (thermolabile) catalase. On the other hand, any identification strategy aiming to go beyond the simple separation of MTB from other mycobacteria entails the use of complex identification techniques capable of addressing a minimum of 20 differentiating features.

Species identification, while one of the steps in the microbiological diagnosis of TB, is of less importance in comparison with culture and particularly, smear microscopy. In countries with high and medium TB prevalence rates, about 99 % of positive smears can be attributed to MTB.

In low- and middle-income countries, cases caused by environmental mycobacteria are so infrequent that they justify the existence of only one laboratory capable of identifying mycobacteria with biochemical tests per country. Often, in countries where culture is indicated, it is only necessary to identify MTB since this is simpler than having to identify the rest of the environmental mycobacteria.

On the other hand, MTB protein 64 (MPT 64) is specific to the MTBC, including MTB, *M. africanum*, *M. bovis*, and some—although not all—*M. bovis* BCG substrains. MPT 64 is an MTBC-specific antigen secreted during bacterial growth and an excellent antigen for MTBC identification. It is an immunochromatography test (ICT) using mouse monoclonal anti-MPT64 to easily differentiate MTBC from non-TB mycobacteria (NTM) [6].

3.3.1 Reporting Results

Laboratories should report initial positive smears, positive MTB cultures, and positive NAAT results within 24 h by telephone or fax to the primary health care provider and health department. Out-of-state laboratories that receive referral specimens must contact the health care provider and health department in the patient's state of origin.

By law and regulation, a case of active TB must be reported to the local health department. Reporting is essential for the TB control program to take action at the local, state, and national levels and to understand the magnitude and distribution of the TB problem.

Reporting renders the health department's resources available to assist the physician in the proper management of the case. Public health services are also available for epidemiological evaluation, including the identification and the examination of source cases and contacts.

3.3.2 In Vitro MTB Drug Susceptibility Testing

In all patients, the initial MTB isolate should be tested for resistance to first-line anti-TB drugs: isoniazid, rifampin, ethambutol, and pyrazinamide. The results of DST should guide clinicians in choosing the appropriate drugs for every patient.

MTB susceptibility testing can be carried out directly with a specimen concentrate if abundant acid-fast bacilli are observed under the microscope (direct method) or in a bacterial culture during the exponential growth phase (indirect method). The standardized techniques for in vitro sensitivity testing are the Canetti proportions and multiple dilutions technique, the Meissner absolute concentration method, and the Mitchison resistance level procedure; all cultures should be performed in Löwenstein–Jensen medium.

Susceptibility test results in LJ medium are available in 4–5 weeks, vs. 2–4 weeks in semi-synthetic media (Middlebrook 7H10 or 7H11) and 5–8 days if using the radiometric technique. These susceptibility tests require these time periods because they are conducted on the positive culture and not on the direct sample (direct method).

The laboratory must inform the clinician on the level of growth occurring in the medium with antituberculous drugs in comparison with an untreated control medium. When susceptibility testing is performed correctly, the control will contain countable colonies.

Here again, the indication to perform the technique will depend on the available resources and the extent of the disease in a specific area. Thus, in rich countries where there are few TB cases and readily available second-line drugs, susceptibility testing should be obtained in at least all patients with a poor microbiological course and in cases of treatment failure, relapse, or treatment default.

Susceptibility testing has an important practical limitation in poor countries, since the information obtained tends to be delayed, is not always reliable and is based on a group of patients (re-treated patients) that constitute a less important epidemiological priority (priority is usually given to initial cases with smear-positive results). The recommended practice is to conduct DST in LJ medium.

Rapid, broth-based systems should be used to identify drug resistance as early as possible in order to ensure appropriate treatment, and susceptibility results should be promptly forwarded to the physician and health department.

Second-line DST should only be conducted in reference laboratories and generally is limited to specimens from patients with the following characteristics:

(1) Prior TB treatment; (2) Contact with a patient with known anti-TB drug resistance; and (3) Demonstrated resistance to first-line anti-TB drugs.

A patient is diagnosed with multidrug-resistant TB (MDR TB) if the organisms are resistant to at least isoniazid and rifampin, the two most potent first-line anti-TB drugs. A patient is diagnosed with extensively drug-resistant TB (XDR TB) if the TB isolate is resistant to isoniazid and rifampin, any fluoroquinolone, and at least one of three injectable second-line drugs (i.e., amikacin, kanamycin, or capreomycin).

3.3.3 Diagnosis of Active TB

PTB

Obtain a chest X-ray; if it suggests TB, order further tests. (1) Submit at least three sputum samples (including one early morning sample) for culture and microscopy. (2) If possible, samples should be spontaneously produced. If not possible: (a) in adults, use induction of sputum or resort to bronchoscopy and lavage and (b) in children, consider induction of sputum if this can be done safely, or gastric washings; if not, (3) Obtain samples before initiating treatment if possible, or within 7 days of treatment administration.

Innovations such as fluorescent staining and LED microscopes have helped to optimize the yield of sputum for smear examination. Mycobacterial sputum culture is considered to be the "gold standard" for the diagnosis of TB and it also facilitates DST.

3.3.4 Volatile Organic Compounds

Although several diagnostic measures have been developed for TB diagnosis, most of these techniques are expensive, e.g., immunological tests using antigens, DNA analysis, or specific culture conditions, especially in developing countries with the highest burden of TB. Thus, the analyses of volatile organic compounds (VOCs) in breath may contain biomarkers of active PTB; they are obtained from breath samples of potentially infected individuals (products of oxidative stress), or from the bacteria per se (metabolites of MTB); these methods have been recently proposed for the rapid diagnosis of TB. Further study is required to determine the metabolic pathways involved and assess whether these compounds are produced during in vivo mycobacterial growth.

Microscopy and culture are still the mainstay of laboratory diagnosis of TB but there is still an urgent need for better diagnostic tools, especially in high-burden countries. An ideal diagnostic test should be sensitive and specific for active pulmonary TB, as well as quick, cost-effective, noninvasive, and suitable for use in developing countries. A breath test might rationally fulfill these requirements because *Mycobacteria* manufacture and release unique volatile VOCs as metabolites with distinctive odors when cultured in vitro.

3.3.5 Breath Sample Collection

A portable breath collection apparatus (BCA) is employed to capture the VOCs in 1.0 L breath and 1.0 L room air in separate sorbent traps. The geometry of the breath reservoir ensures that the sample contains >99 % alveolar gases.

3.3.6   Breath Sample Analysis

VOCs captured in the sorbent traps are analyzed in the laboratory by automated thermal desorption, gas chromatography, and mass spectroscopy (ATD/GC/MS). In order to quantify peak areas and to control for drift in instrument performance, an internal standard is run with every chromatographic assay of breath and air (0.25 mL 2 ppm 1-bromo-4-fluorobenzene, Supelco, Bellefonte, PA). MS is expensive because it requires sophisticated pumping and sealing technology to generate and maintain a high vacuum; an instrument modified for breath VOC assays may cost approximately $200,000.

Breath testing is rapid, noninvasive, and completely safe. It could potentially be employed as a primary screening tool in populations at increased risk of TB. For this reason, an important objective of future research will be to adapt the breath test to a less expensive analytical platform but that still delivers results within minutes at the point-of-care. This objective has yet to be fulfilled, although its feasibility is supported by recent advances in breath testing technology. Breath tests detect volatile biomarkers of active PTB and have potential applications in clinical practice.

Significant qualitative and quantitative differences (number of compounds as well as probable higher emission rates) have been detected in the bouquet of volatile metabolites in MTB, NTM, and other bacteria. This behavior differentiates MTB from other mycobacteria, which generally produce fewer compounds and in seemingly lower quantities. Knowledge of the production of volatile gases by MTB can facilitate the rational design of alternative and faster diagnostic measures.

Further studies are required to elucidate the metabolic pathways involved and assess whether these compounds are produced during mycobacterial growth in vivo.

Bacilloscopy (sputum or other samples) remains the most readily available test to establish a microbiological diagnosis but other more sensitive methods identifying MTB, particularly rapid molecular tests, are quickly gaining acceptance due to their performance and applicability. Table 3.2 succinctly summarizes the evidence-based performance of different TB diagnostic tests [5, 10, 11].

3.3.7   Active Non-PTB

If non-respiratory TB is a diagnostic possibility, part or all of any of the following samples should be placed in a dry container (and not all placed in formalin) and submitted for TB culture:

1. Lymph node biopsy
2. Pus aspirated from lymph nodes
3. Pleural biopsy
4. Any surgical sample submitted for routine culture
5. Any radiologically obtained sample submitted for routine culture
6. Histology sample

Table 3.2 WHO-approved microbiological tests for TB

Test	Site	Principal findings in systematic reviews
Active TB diagnosis		
Sputum smear microscopy	Pulmonary	• Fluorescence microscopy is 10 % more sensitive than conventional microscopy, with similar specificity and time efficiency • The examination of two smears obtained on the same day is as precise as a standard smear, with the same sensitivity (63 %) and specificity (98 %)
Nucleic acid amplification tests (NAAT) [no-MTB Xpert/RIF]	Pulmonary TB and (EPTB)	• Standard, commercial NAAT has a high specificity and positive predictive value, but lower sensitivity and negative predictive value in all forms of TB, particularly in those with negative bacilloscopies and in EPTB
Xpert MTB/RIF	PTB, EPTB, and RIF resistance	• Xpert MTB/RIF used as an initial diagnostic test for detection of MTB and RIF resistance; it is sensitive and specific. When used instead of bacilloscopy, its sensitivity is 88 % and its specificity is 98 %; detecting RIF resistance, its sensitivity is 94 % and its specificity is 98 %
Automated liquid culture and species identification based on MPT64	PTB, EPTB; speciation	• Automated liquid cultures are more sensitive than solid cultures and the time to detection is quicker • Rapid Immunochromatographic tests (ICT) based on MPT64 for species identification are highly sensitive and specific

7. Aspiration sample
8. Autopsy sample

Discuss the advantages and disadvantages of biopsy and needle aspiration with the patient. Obtain a chest X-ray to exclude coexisting PTB in all patients with non-PTB. Microbiology staff should routinely perform TB cultures on the above samples (even if it is not requested).

Rapid Diagnostic Tests

Rapid diagnostic tests for MTBC (*M. tuberculosis*, *M. bovis*, *M. africanum*) on primary specimens should be used only if: (1) Rapid confirmation of TB in a sputum smear-positive patient would modify their care, or (2) Before conducting a large contact-tracing initiative.

If clinical signs and other laboratory findings are consistent with TB meningitis, begin treatment even if a rapid diagnostic test is negative. If risk assessment suggests that a patient has MDR TB: (1) Conduct rapid diagnostic tests for rifampicin resistance and (2) Initiate infection control measures and treatment of MDR TB while awaiting the results.

Clinicians should still consider a diagnosis of non-respiratory TB if rapid diagnostic tests are negative in for example, pleural fluid, cerebrospinal fluid, and urine.

1. Clinical signs and other laboratory findings consistent with TB meningitis should prompt treatment initiation even if a rapid diagnostic test is negative since the infection's potential consequences are dire.
2. Before conducting a large contact-tracing initiative (for example, in a school or hospital), the species of *Mycobacterium* should be confirmed to be MTBC by rapid diagnostic tests, on microscopy or in culture positive material. Clinical judgment should be used if tests are inconclusive or delayed.
3. If risk assessment suggests that a patient has multidrug-resistant (MDR) TB, rapid diagnostic tests should be conducted for rifampicin resistance, and infection control measures and treatment of MDR TB should be initiated, pending the test results.
4. Rapid diagnostic tests for MTBC identification should be conducted on biopsy material only if: the entire sample has been inappropriately placed in formalin *and* AFB are visible on microscopy.
5. Clinical samples should ideally be submitted for culture by automated liquid methods, bearing in mind that laboratories need a certain level of throughput to maintain quality control.

References

1. American Thoracic Society. Diagnostic standards and classification of tuberculosis in adults and children. Am J Respir Crit Care Med. 2000;161:1376–95.
2. Bernard C, Wichlacz C, Rigoreau M, Sorhouet S, Dagiral R, Jarlier V, et al. Evaluation of the Fluo-RAL module for detection of tuberculous and nontuberculous acid-fast bacilli by fluorescence microscopy. J Clin Microbiol. 2013;51:3469–70.
3. Davis JL, Cattamanchi A, Cuevas LE, et al. Diagnostic accuracy of same-day microscopy versus standard microscopy for pulmonary tuberculosis: a systematic review and meta-analysis. Lancet Infect Dis. 2013;13:147–54.
4. García-Elorriaga G, Carrillo-Montes G, del Rey-Pineda G, González-Bonilla C. Detection of *Mycobacterium tuberculosis* from respiratory simples with the liquid culture system MB/BacT and verified by PCR. Rev Invest Clin. 2006;58:573–9.
5. Ling DI, Flores LL, Riley LW, Pai M. Commercial nucleic-acid amplification tests for diagnosis of pulmonary tuberculosis in respiratory specimens: meta-analysis and meta-regression. PLoS One. 2008;3:e1536.
6. Maurya AK, Nag VL, Kant S, Kushwaha RA, Kumar M, Mishra V, et al. Evaluation of an immunochromatographic test for discrimination between Mycobacterium tuberculosis complex & non tuberculous mycobacteria in clinical isolates from extra-pulmonary tuberculosis. Indian J Med Res. 2012;135:901–6.
7. National Institute for Health and Clinical Excellence. Tuberculosis: clinical diagnosis and management of tuberculosis, and measures for its prevention and control. NICE Clinical Guideline 117. London: NICE; 2011.
8. Pfyffer GE, Palicova F. Chapter 28: *Mycobacterium*: general characteristics, laboratory detection, and staining procedures. In: Versalovic J, editor. Manual of clinical microbiology. 10th ed. Washington: ASM Press; 2011.

 9. Steingart KR, Henry M, Ng V, et al. Fluorescence versus conventional sputum smear micros-copy for tuberculosis: a systematic review. Lancet Infect Dis. 2006;6:570–81.
10. Steingart KR, Schiller I, Horne DJ, et al. Xpert® MTB/RIF assay for pulmonary tuberculosis and rifampicin resistance in adults (review). Cochrane Database Syst Rev. 2014;1, CD009593. doi:10.1002/14651858.CD009593.pub2.
11. TB CARE I. International standards for tuberculosis care. 3rd ed. The Hague: TB CARE I; 2014.

Chapter 4
Molecular Diagnosis

Abstract Isothermal DNA amplification techniques are simple, rapid, cost-effective and they have the equivalent specificity and sensitivity to PCR, thus enabling point-of-care diagnostics for pathogen detection, single nucleotide polymorphism analysis, biomarker detection, etc. Future work in this area should focus on microfluidic adaptations of these techniques with integrated sample preparation and further reduction of cost and result turnover time. Effective amplification of large genomic segments by isothermal methods results in unbiased whole genome amplification from a few homogeneous cells, serving as high-quality single-cell genetic material. This material can then be used for archiving and downstream analysis such as genotyping, comparative genomic hybridization, and single-cell genomics. Immunoassays coupled with isothermal DNA amplification have improved sensitivities compared with traditional enzyme-amplified immunoassays, and improved dynamic range and reproducibility compared to immuno-PCR.

Highly specific isothermal amplification reactions (LAMP, HDA) have allowed the development of assays in which the nonspecific detection of amplicon accumulation is sufficient to indicate the presence of an initial template, allowing real-time fluorescent detection with an intercalating DNA dye such as EvaGreen or SYBR Green I.

Keywords Isothermal NAAT • Molecular diagnosis • NAAT • PCR

4.1 Introduction

In spite of the many advances in the fight against TB, 8.6 million individuals developed TB in 2012 and 1.3 million died as a result. Many of these deaths could have been avoided if diagnosis and treatment had been timely, but approximately a third of TB patients have no access to appropriate diagnostic measures [11].

Drug-resistant TB, including multidrug-resistant TB (MDR TB, defined as resistance to at least isoniazid and rifampicin, the two most important anti-TB first-line drugs) and extensively drug-resistant TB (XDR-TB, defined as MDR-TB plus resistance to any fluoroquinolone such as oxifloxacin or moxifloxacin and at least one of three injectable second-line drugs—amikacin, capreomycin, or kanamycin), have

© Springer International Publishing Switzerland 2015   35
G. García-Elorriaga, G. del Rey-Pineda, *Practical and Laboratory Diagnosis of Tuberculosis*, SpringerBriefs in Microbiology,
DOI 10.1007/978-3-319-20478-9_4

become a serious threat to world health. In 2012, about 450,000 individuals developed MDR-TB and 170,000 are estimated to have died from MDR-TB.

The basic principles underlying the International Standards for Tuberculosis Care (ISTC) have not changed. Case detection and curative therapy are still the mainstay of TB care and its control, but the fundamental responsibility of health care providers in order to ensure ending treatment remains unchanged [3].

Molecular diagnostic assays are based on the detection of specific nucleic acid molecules contained in microorganisms. Identification of the presence of a microorganism's nucleic acids in a patient sample allows for the positive identification of that microorganism.

Molecular techniques are well suited to the identification of infectious microorganisms in human samples since every microorganism contains specific DNA or RNA sequences that are unique to that organism. By detecting these unique sequences, the microorganism can be specifically identified.

The main advantages of amplification assays for the detection of microorganisms are their higher sensitivity and faster turnover rate compared with many current microbiology techniques. Amplification assays are theoretically capable of detecting as little as one microorganism in a sample.

Over the past decades, major advances have been made in molecular diagnostic testing in the realm of infectious disease. Today, molecular diagnostic tests are playing an increasingly important role in the clinical microbiology laboratory because they provide superior performance and a faster turnover rate than traditional tests. Despite these advances, culture remains the gold standard for microbial identification and subsequent phenotypic differentiation of the causative pathogen [6].

DNA replication is an indispensable process in biological systems. This process generates copies of the genetic instructions that are required to maintain life; thus, its strategies and components have been evolutionarily optimized. Sequence-based amplification of specific genes has many applications in molecular biology research and medical diagnostics. Currently, there are two main strategies used for the amplification of a defined sequence of nucleic acids: the polymerase chain reaction (PCR) and isothermal amplification. PCR is one of the most widely used DNA amplification techniques, it relies on instrument-based thermal cycling that denatures the DNA template, followed by the annealing of primers at specific sites in the denatured template and extension of the primers by a thermostable DNA polymerase that exponentially increases the amount of DNA. Isothermal amplification of DNA requires the same three steps, but reagents and conditions are specifically chosen to allow for the amplification of the DNA at a defined temperature, after the initial high temperature incubation required for denaturing the DNA template.

In order to circumvent the limitations of traditional PCR in the amplification stages at molecular diagnostics point-of-care testing (POCT) sites, recent research has turned toward isothermal methods for nucleic acid amplification. Isothermal amplification techniques rely on enzymes that separate strands that would otherwise require repeated heating to obtain.

4.2   Nucleic Acid Amplification Test

NAAT is a valuable molecular tool not only in basic research but in practical fields as well, such as in the development of clinical medicine, the diagnosis of infectious disease, gene cloning, and industrial quality control, among others. Several amplification methods have already been developed such as the PCR and the loop-mediated isothermal amplification procedure (LAMP), strand displacement amplification (SDA), and rolling circle amplification (RCA).

Key requirements for any sputum-based nucleic acid amplification test (NAAT) to be performed at a TB POCT, in peripheral microscopy centers of high-burden countries include:

1. Precision:

   – The assay must be more sensitive than sputum microscopy and ideally, as sensitive as the Xpert® MTB/rifampin (RIF) assay (GeneXpert) in the detection of pulmonary TB (PTB).
   – It must be at least as specific as bacilloscopy and the Xpert® MTB/RIF assay in PTB detection.
   – Added capacity is desirable to detect drug resistance or it may be an add-on (reflex) test if not integrated in the initial detection kit.

2. Operational aspects:

   – It should not be more complex than bacilloscopy.
   – Basic laboratory technicians with minimum training should be able to perform the NAAT.
   – Manual precision steps should be kept at a minimum (similar to the GeneXpert), especially in terms of sample processing.

3. Cost:

   – It may be more costly than bacilloscopy but less expensive than the GeneXpert.

4. The main advantage of molecular assays is their speediness: one can identify MTB and gene mutations associated to anti-TB drug resistance and decrease the time period until resistance detection, to 1 or 2 days. Compared to the gold standard (culture and conventional DST), one of its disadvantages is that molecular methods are unable to determine the proportion of drug-resistant bacteria in the sample.

The presented evidence and expert opinion underscore the fact that rapid molecular assays are valuable in TB identification and the detection of drug resistance, especially in patients suspected of harboring PTB and a positive smear (including those HIV-positive), and in the diagnosis of some forms of extrapulmonary TB (EPTB). These methods, however, should not replace standard diagnostic methods (including clinical expertise, microbiological, and imaging methods) and conventional drug sensitivity testing for the diagnosis of active TB and drug resistance (RIF and isoniazid [INH]) in suspected PTB and EPTB cases.

4.2.1   PCR

The advent of PCR revolutionized genetics and molecular diagnostics, by providing a simple and elegant method for nucleic acid amplification using thermostable polymerase enzymes and cyclical heating and cooling to obtain strand separation and annealing, respectively. This powerful technology has been well characterized and is widely employed for molecular diagnosis, biomedical, and life science research.

4.2.2   Line Probe Assays (LPAs) (INNO-LiPA Rif TB Assay, MTBDRsl)

Alternative molecular methods to detect drug sensitivity (DST) include assays with commercial probes, INNO-LiPA Rif.TB (Innogenetics, Ghent, Belgium), and the Genotype® MTBDRplus assay (Hain LifeScience GmbH, Nehren, Germany). They identify the MTB Complex, and the targets of the INNO-LiPA Rif.TB assay are common mutations in the rpoB gene, associated to RIF resistance, while the targets of the GenoType® MTBDRplus assay also include common mutations in the katG and INHA genes, associated to INH resistance, rpoB mutations and it can also detect the MTB Complex.

4.2.3   GeneXpert

Recently developed automated nucleic acid amplification technology can simultaneously and rapidly detect the MTB Complex and RIF resistance; it is widely recognized as an MDR-TB marker (Cepheid GeneXpert MTB/RIF), it is the most advanced and operates in Real Time (RT)-PCR [10].

The Xpert® MTB/RIF (GeneXpert) test (Cepheid Inc., Sunnyvale, CA; http://www.cepheid.com) was endorsed by the World Health Organization (WHO) in their published guide (2011). This new guide has broadened the recommended use of the GeneXpert, even for the diagnosis of childhood and EPTB; it includes an additional recommendation on the use of the GeneXpert as an initial diagnostic test in all individuals suspected of having PTB. The WHO policy recommended that GeneXpert should also be used as an initial diagnostic test in individuals suspected of MDR or HIV-associated TB.

The detection limit of the GeneXpert (in a reproducible manner based on negative samples and a 95 % confidence interval) is five copies of the genomic DNA purified by the reaction or 131 MTB colony-forming units per mL of sputum. In comparison, the identification of the TB bacillus by microscopy requires at least 10,000 bacilli/mL of sputum. The GeneXpert detects both live and dead bacteria.

A Cochrane systematic review by Steingart and colleagues included 18 published studies on the accuracy of the GeneXpert for PTB detection and rifampicin resistance in adults. The meta-analysis showed the test was highly accurate, reinforcing the WHO's endorsement of this technology. Since the publication of Steingart's original Cochrane Review [14], nine additional new studies have been added. Of the 27 studies, two were international and multicentric and conducted in five and six study centers, respectively. The findings in the update were consistent with the previous reports [8].

When used as an initial test *in lieu* of smear microscopy (15 studies, 7517 participants), GeneXpert achieved a pooled sensitivity of 88 % (95 % credible interval [CrI] 83–92 %) and a pooled specificity of 98 % (95 % CrI 97–99 %). The pooled sensitivity was 98 % (95 % CrI 97–99 %) for smear-positive, culture-positive TB and 68 % (95 % CrI 59–75 %) for smear-negative, culture-positive TB (15 studies); the pooled sensitivity was 80 % (95 % CrI 67–88 %) in individuals with HIV and 89 % (95 % CrI 81–94 %) in those without HIV infection (four studies).

In the detection of rifampicin resistance (11 studies, 2340 participants), GeneXpert achieved a pooled sensitivity of 94 % (95 % CrI 87–97 %) and pooled specificity of 98 % (95 % CrI 97–99 %). Thus, a GeneXpert result that is positive for rifampicin resistance should be carefully interpreted and consideration should be taken of the risk of MDR-TB in a given patient and the expected prevalence of MDR-TB in a given setting.

Future studies should determine the GeneXpert's diagnostic accuracy in peripheral laboratories and clinical settings such as primary care centers, TB detection centers, antiretroviral clinics, particularly in those in which the test is performed on-site. Systematic reviews have been conducted on GeneXpert in the diagnosis of EPTB and pediatric TB; and the findings described in the updated WHO policy declare the use of GeneXpert.

4.2.4   Policy Updates

The WHO endorsed the GeneXpert in 2010 and published an accompanying policy to guide its use in 2011. In October 2013, the WHO issued updated recommendations on the use of the GeneXpert.

This new policy broadens the recommended use of the GeneXpert, to include the diagnosis of childhood TB and EPTB, and additionally recommends it as the initial diagnostic test in all individuals presumed to have PTB.

The revised WHO recommendations on the use of the GeneXpert for the diagnosis of PTB and rifampicin (RIF) resistance in adults and children are as follows:

1. The GeneXpert should be used rather than conventional microscopy, culture, and drug susceptibility testing as the initial diagnostic test in adults presumed to have MDR-TB or HIV-associated TB (strong recommendation, high-quality evidence).

2. The GeneXpert should be used rather than conventional microscopy, culture, and drug susceptibility testing as the initial diagnostic test in children presumed to have MDR-TB or HIV-associated TB (strong recommendation, very low-quality evidence).
3. The GeneXpert may be used rather than conventional microscopy and culture as the initial diagnostic test in all adults presumed to have TB (conditional recommendation acknowledging resource implications, high-quality evidence).
4. The GeneXpert may be used rather than conventional microscopy and culture as the initial diagnostic test in all children presumed to have TB (conditional recommendation acknowledging resource implications, very low-quality evidence).
5. The GeneXpert may be used as a follow-up test to microscopy in adults presumed to have TB but not at risk for MDR-TB or HIV-associated TB, especially in subsequent testing of smear-negative specimens (conditional recommendation acknowledging resource implications, high-quality evidence).

The revised recommendations on the use of the GeneXpert for the diagnosis of EPTB are as follows:

1. The GeneXpert should be preferably used instead of conventional microscopy and culture as the initial diagnostic test when testing cerebrospinal fluid specimens from patients presumed to have TB meningitis (strong recommendation given the urgency of a rapid diagnosis, very low quality of evidence).
2. The GeneXpert may be used as a replacement test for usual practices (including conventional microscopy, culture, and/or histopathology) to test specific, non-respiratory specimens (lymph nodes and other tissues) from patients presumed to have EPTB (conditional recommendation, very low quality of evidence).

Evidence from systematic reviews included in the revised recommendations, is published in the report of the Expert Group Meeting and in the policy itself. These revised recommendations also are incorporated into the third edition of the International Standards for TB Care (www.istcweb.org).

4.2.5   Implementation of Existing Technologies

However in most countries, implementation of the GeneXpert as a TB diagnostic method in all individuals with signs or symptoms of TB, would cost more than the use of conventional diagnostic methods (including smear microscopy, chest X-ray follow-up in patients with a negative bacilloscopy).

According to the WHO, as of September 30, 2013, a total of 1843 GeneXpert® instruments and 4 214 990 GeneXpert cartridges have been procured worldwide by the public sector in 95 of the 145 countries eligible for concessional pricing. Updated quarterly sales figures are publicly available via the WHO website to monitor the GeneXpert rollout.

As of March 31, 2013, Brazil had purchased 34,260 cartridges. Based on a pilot roll-out of the Xpert® MTB/RIF in two municipalities, Brazil has made plans to replace all diagnostic smear microscopy with the Xpert® MTB/RIF assay.

In June 2012, UNITAID, the Bill & Melinda Gates Foundation, the United States Agency for International Development (USAID) and the United States President's Emergency Plan for AIDS Relief (PEPFAR), announced an agreement with Cepheid Inc. to decrease the cost of the test to US$9.98 per cartridge (from US$16.86), a price that will remain in place until 2022. This purchase price is applicable to over 145 purchasers in low- and middle-income countries. In addition to these global developments, efforts are underway to promote the introduction of the GeneXpert into the private sector in high-burden countries such as Bangladesh, India, Indonesia, and Pakistan. Currently, these purchasers are excluded from accessing the negotiated price of US$9.98 per test cartridge [9].

The UNITAID Tuberculosis Diagnostics Landscape is published as part of a broad and continuous effort to understand the technology perspective in terms of TB diagnosis. The first edition of the UNITAID Tuberculosis Diagnostics Landscape was published in July 2012; it established TB diagnostic techniques including bacilloscopy and culture as well as emerging tools such as NAAT. A semiannual update was published in December 2012. A second edition was published in July 2013 [1]. These documents are available at: http://www.unitaid.eu/en/resources/publications/technical-informes.

The purpose of this document is to put the Xpert® MTB/RIF assay in practice and define the characteristics of the next generation of molecular tools that may replace bacilloscopy.

TB REACH, an initiative of the Stop TB Partnership supported by the Government of Canada, promotes new ways of detecting and treating TB cases. In its first wave of grants, TB REACH supported a project that used the GeneXpert in a mobile van in Tanzania even prior to WHO endorsement of the technology.

4.2.6 Planned Technology Refinements of GeneXpert

Recent technology developments of the GeneXpert include:

1. Development of ten color channel detection with high-resolution melt capability. This will expand the multiplexing capability of the existing installed base of GeneXpert systems without the need for hardware upgrades or replacement.
2. Remote calibration was released late in 2012 and is now used in over 40 countries. Remote calibration allows for GeneXpert modules to be calibrated without the need for an expensive technician service call or visit; over 90 % of modules can be calibrated over the Internet.
3. GeneXpert Data Management initiatives to allow for real-time aggregation of de-identified geo-positioned test data. Proof-of-concept studies are being conducted in South Africa and will soon be in the United States, in order to assess the ability to monitor and track disease incidence and drug resistance.

4.3   Other Isothermal NAATs

NAAT promises rapid, sensitive and specific diagnosis of infectious disease. The next generation of diagnostic devices will search for the genetic determinants of such conditions at the point-of-care, affording clinicians a prompt and reliable diagnosis with which to guide an effective therapeutic approach.

The discovery of the structure of DNA, the elucidation of the molecular mechanisms of genetics, and the development of the PCR have given rise to powerful methods for the diagnosis of infectious disease. Despite these advances, the gold standard for microbial identification is still culture and subsequent phenotypic differentiation of the causative pathogen.

NAAT techniques allow further clinically pertinent information to be garnered from either patient or pathogen. Antimicrobial resistance, virulence biomarkers and highly specific typing can be rapidly identified, allowing optimal therapy and therapeutic interventions to proceed without delay.

A role in which NAAT will prove especially powerful, is the rapid differentiation of pathogens whose prodromal presentation is similar, but may require significantly distinct therapeutic strategies. The goal of NAAT, either in the laboratory or POC, is to identify and potentially quantify specific nucleic acid sequences from clinical samples, as markers of the presence of an infectious process.

These methods involve sequential steps. First, the amplification—typically by PCR—of NA isolated from a clinical sample, followed by the detection of the amplification product.

4.3.1   Transcription Mediated Amplification/Nucleic Acid Sequence Based Amplification

Nucleic acid sequence-based amplification (NASBA), also known as transcription-mediated amplification (TMA), is a sensitive amplification system based on transcription (TAS) and used in the specific replication of nucleic acids in vitro.

NASBA characteristics: (1) One-step isothermal amplification reaction at 41 °C; (2) Especially recommended for RNA analytes due to reverse transcriptase (RT) integration in the amplification process; (3) The product of single-chain RNA is an ideal target for detection with different methods including probe hybridization.

The isothermal nucleic acid amplification of RNA is a result of the simultaneous action of the avian myeloblastosis virus reverse transcriptase (AMVRT), the T7 RNA polymerase, and RNase H. NASBA was developed in the early 1990s to amplify nucleic acids without the need for a thermal cycler.

TMA can be used to target both RNA and DNA. TMA has several other differences in comparison to PCR: (1) TMA is isothermal. A water bath or heat block is used instead of a thermal cycler; (2) TMA produces an RNA amplicon rather than a DNA amplicon. Since RNA is more labile than DNA in the laboratory environment, it decreases the possibility of carry-over contamination; (3) TMA produces 100–1000 copies per cycle in contrast to PCR that produces only two copies per cycle.

The product of the NASBA reaction is mainly single-chain RNA. Its detection used to require a very labor-intensive procedure requiring agarose gel electrophoresis stained with ethidium bromide (EtBr). Currently, NASBA products are detected at the end of amplification with electrochemiluminescence (ECL). Another detection method used in research and NASBA qualitative reactions, is the enzyme-linked gel assay (ELGA). This is a very recent and elegant method designed to detect NASBA products with molecular probes.

NASBA offers several advantages over other mRNA amplification methods. Amplification of nucleic acid sequences with over 109 copies can be performed in only 90 min by the simultaneous action of the three enzymes. (1) NASBA is an isothermal reaction conducted at 41 °C, thus avoiding the need for a thermal cycler and promoting POCT. (2) One of NASBA's main advantages is the production of single-stranded RNA amplicons that can be directly used in another amplification round or for detection without the need for denaturing or strand separation. (3) NASBA was specifically designed to detect RNA.

NASBA also has some disadvantages. (1) RNA integrity is the main concern when performing NASBA, as well as RT-PCR and other RNA amplification procedures. (2) Although the amplification reaction per se is isothermal at 41 °C, a single melting stage before the amplification reaction is required to permit alignment of the initiators with the target. (3) Moreover, since the reactions' specificity is dependent on thermolabile enzymes, the reaction temperature cannot exceed 42 °C, without compromising the results.

For POCT adaptation, the low incubation temperature (41 °C) is desirable in order to decrease power consumption and thermal control complexity, but may result in a low stringency reaction environment and nonspecific amplification; robust primer design and assay evaluation is therefore crucial. Furthermore, an initial 95 °C strand separation step is required if dsDNA is targeted while RNA amplification requires a 65 °C step to remove secondary structures.

A NASBA assay was designed based on a highly conserved mycobacterial 16S rRNA. A species-specific probe was used to detect MTB amplicons in an ELGA-based detection system. Nucleic acids obtained from 200 bacteria resulted in a positive signal. The assay correctly identified 32 MTB strains from different parts of the world [2].

4.3.2   Simple Method for Amplifying RNA Targets

This method incorporates the advantages of a microfluidic technique and avoids the limitations of NASBA (including amplified RNA segment inhibition by the secondary structure and binding sites limited to 100–250 nucleotides). Simple method for amplifying RNA targets (SMART) fulfills this objective by incorporating a binding stage and a microfluid separation stage, allowing both hybridization sites to be located anywhere along the RNA target [5]. Further, the incorporation of a probe designed from ssDNA allows the user to choose an amplifiable probe and primer sequences that decrease the secondary structure and optimizes the kinetic reaction.

The SMART method uses two probes that bind to specific sequences in the RNA target, allowing the investigator to test many characteristics of a pathogen such as sub-type, drug resistance, and strain.

4.3.3 Recombinase Polymerase Amplification

Recombinase polymerase amplification (RPA) replaces PCR as a rapid, isothermal, enzymatic process.

RPA is a single tube, single temperature (37–42 °C) amplification method. The key to the amplification process is the formation of a recombinase filament, a complex combining a target-specific primer and a recombinase enzyme. Products can be detected in real time with fluorescent probes or endpoint methods such as lateral flow strips or gel electrophoresis.

The favorable thermal requirements, the procedure's simplicity and very rapid amplification (20–40 min), make this cutting-edge technology ideal for integration among POCT techniques.

Regardless of the amplification method, a key step in selective amplification of target DNA is primer/probe annealing, which is usually kinetically driven using high concentrations of the primer/probe. For this purpose, the annealing temperature (~65 °C) is preset.

Benefits of RPA: (1) Speed—Results in 10–15 min; (2) Sensible—Detection of a single molecule; and (3) Low cost—Minimal or no hardware is necessary.

4.3.4 Helicase-Dependent Amplification

Helicase-dependent amplification (HDA) is an elegant method for DNA amplification, employing helicase enzymes for strand separation, as opposed to the thermal strand separation used in PCR. The developers have created a simple and not unfamiliar reaction model of isothermal amplification that more closely resembles in vivo DNA replication.

The process initiates with double-stranded DNA (dsDNA) unwinding by helicase and to which forward and reverse primers can bind, followed by polymerase-mediated elongation. After elongation, helicase can act again on the freshly synthesized dsDNA and the cycle is asynchronously repeated, with amplification kinetics similar to those of classic PCR, at 60–65 °C and without further temperature modification steps. The existing HDA protocol is typically 60–120 min for small numbers of copy targets.

The HDA method is a simple reaction requiring a couple of primers, a reaction buffer, and an enzyme mixture similar to that used in PCR. The reaction is unlike other isothermal methods that have more complex primer and enzyme requirements and that actually need to undergo two temperature changes.

A microfluid chip can perform HDA on samples with live bacteria and can detect 10 CFU. Preparation of the sample with the chip-based technique in conjunction with isothermal amplification is an attractive option due to its potential worldwide applications, since the need for handling for flow control and temperature changes is minimal or eliminated.

The primary appeal of HDA for POCT applications is the relative simplicity of the reaction. A single set of primers, two enzymes (three for reverse transcription-HDA) and compatibility with existing fluorescent detection chemicals make this method an easy alternative to those familiar with PCR.

4.3.5   Rolling Circle Amplification

This is a method to amplify short DNA molecules immobilized on a hard surface. It is based on a solid phase rolling circle replication reaction known as RCA.

The DNA sequences are specific to the RCA targets by hybridization and the DNA ligase reaction. The circular product is then used as an RCA template.

RCA amplifies the signal on a solid phase using small DNA probes immobilized on a solid surface. Since the probes are fixed to a solid surface, different groups of probes can be fixed in known positions and generate a matrix of tests capable of screening many sequences in a parallel manner.

RCA technology is sufficiently versatile to allow the use of different DNA recognition methods and detect the degree of amplification of the immobilized DNA probes. Although the incorporation of radiolabeled nucleotides has been used to detect signal amplification in initial tests, other approaches can be employed including the direct incorporation of fluorescence-labeled nucleotides in the amplification process or probe hybridization in the amplified product.

RCA combines the specificity of hybridization and ligation with amplification, thus obtaining a solid phase method to detect a DNA target. It allows the recognition of highly specific DNA sequences and greatly amplifies the signal obtained from small initiator probes.

Low (30–60 °C) temperature requirements make these techniques attractive for POCT and have proven successful when using bacterial targets in traditional laboratory assays.

In addition to the linear RCA kinetics originally described in a ~4-h amplification, more powerful RCA amplification variants have been developed, capable of generating detectable product levels in 30–90 min with a sensitivity as low as ten copies. Referred to as geometric, hyperbranched, ramification, or cascade RCA, these methods make use of secondary primers that target the amplification product.

4.3.6 Ramification-Extension Amplification

This is an alternative amplification method that avoids thermal cycler-related difficulties and makes the assays easier and more practical in the clinical laboratory. It combines the technology of magnetic pearls to isolate the nucleic acid target in samples; it generates functional target-dependent probes and amplifies the isothermal probe; and it is a simple and practical amplification technique.

Nucleic acid targets (DNA or RNA) are isolated with DNA capture probes that have a 3′ complementary sequence to the target and a biotin residue in the 5′ terminus that can interact with paramagnetic pearls covered with streptavidin. After cell lysis with guanidine thiocyanate (GTC) that inactivates pathogens, hybridization of the probe to the target can be completed without previous isolation of the target nucleic acids and thus eliminating the possibility of infections among the laboratory personnel and decreasing the assay's duration.

Ramification-extension amplification (RAM) requires single-stranded circular DNA, a single enzyme, and a couple of initiators. After recognizing the target and C-probe lineal ligation, circular single-stranded DNA (ssDNA) is amplified instead of the target sequences [7].

This C-probe includes three regions: a 5′ region (25 nucleotides); a 3′ region (25 nucleotides) that complements the target sequence in an adjacent position, and a linker region (59 nucleotides) composed of a linking generic RAM initiator. The introduction of ligation in the RAM system increases the assay's specificity since ligation would not be possible unless the 3′ and 5′ termini hybridize with the target in perfect alignment.

After ligation, several forward initiators (initiators–F), reverse initiators (initiators–R), and DNA polymerase are directly added to the reaction.

If an initiator–F and an initiator–R are included in the RAM reaction, growth is exponential. A mathematical formula expressing the reaction has been developed: $2U$, where U is the number of amplification cycles. This formula shows that extension-ramification amplifies exponentially.

This system has many advantages over other techniques:

1. This system uses the properties of DNA polymerase extension and displacement. Therefore, an enzyme can both extend and ramify.
2. Generic extension initiators are used for all target nucleic acids (i.e., all infectious agents); the initiator-binding sequence (binding region) is identical in all circular probes regardless of the hybridization terminus (5′ and 3′). Consequently, the detection of multiple infectious agents in the same sample is possible.
3. Since extension and ramification of the initiator generates larger polymers (>8000 bp), no thermal cycler is required. It is an ideal method for in situ amplification.

This is a novel nucleic acid amplification system with many unique characteristics. It can be easily used in clinical laboratories to detect infectious agents.

4.3.7   Loop-Mediated Isothermal Amplification

In the past few years, the Foundation for Innovative New Diagnostics (FIND) partnered with the Eiken Chemical Company (Eiken), Tokyo, Japan, to develop molecular assays designed to detect several infectious diseases (such as TB) using their LAMP platform. This molecular technique was chosen due to specific characteristics that favor its use in simplified testing systems that could be appropriate in settings with limited resources [4, 13].

Evidence on the manual NAAT, developed by Eiken Chemical Corp. (Japan) and the FIND, was also reviewed in 2012 by a WHO Group of Experts (April 20, 2012). The LAMP platform has been developed as replacement of microscopy.

Among the currently available nucleic acid isothermal methods, LAMP is the most widely researched and has been well characterized; offering significant support during the development process.

There is limited published data on these new molecular tests. Most do not appear to be ready for revision of the WHO policy (maybe in 2–3 years), except for the Eiken Loopamp™ MTBC assay, programmed for review by the WHO in 2014.

The amplification reaction requires four types of initiators that complement six regions of the target gene. Since double-stranded DNA is in a state of dynamic equilibrium at a temperature of about 65 °C, one of the LAMP initiators can hybridize with the complementary sequence of the double-chain DNA target and initiate DNA synthesis with a DNA polymerase with chain displacement activity and finally releasing single-stranded DNA.

LAMP has specific advantages that could foster the development of simple and economical molecular tests that could be used in sputum microscopy. These characteristics make LAMP an attractive platform for the development of NAATs applicable in developing countries.

LAMP is a rapid amplification method employing a strand-displacing *Bacillus stearothermophilus* (Bst) DNA polymerase. This results in an amplification model in which the priming sequence is copied with each round of replication and remains tethered to the previous amplicon, resulting in a concatenated product of alternating sense/antisense repeats of varied length. The 60–65 °C reaction temperature combined with a minimum of four primers makes LAMP a highly specific reaction, allowing an "amplification is detection" format.

The final version of the TB-LAMP includes the following steps:

1. Sample preparation (10–20 min):

   – With a disposable pipette (provided by Eiken), collect 60 μL of sputum from a sputum sample and transfer the collected sputum to a heating tube containing extraction solution.
   – Mix by inverting 3–4 times and place the heating tube in heating block at 90 °C for 5 min to lyse and inactivate mycobacteria.
   – Remove the heating tube from the heating block and allow it to cool down for 2 min.

2. Amplification (40 min):

  - Confirm that the temperature on the digital display on the incubator is 67 °C.
  - Load the reaction tubes into the heating block and begin the reaction.
  - The amplification is stopped automatically after 40 min.

3. Visual detection of fluorescence light from the reaction tube using UV light (0.5–1 min):

  - Transfer the reaction tubes into the fluorescence detector and record the results.
  - Discard the reaction tubes (without opening them) by incineration.

LAMP has been well characterized and widely employed for bacterial pathogen detection. These assays have all been described as performing equally to or better than the equivalent PCR, immunoassay or culture-based diagnostic methods. Analytical sensitivity has been shown to exceed that of equivalent PCR assays with detection limits as low as five copies. These attributes make LAMP well suited for adaptation into microfluidic and POC assays.

4.3.8   Cross-Priming Amplification

Cross-priming amplification (CPA) refers to a class of nucleic acid isothermal amplification reactions requiring multiple initiators and probes, one or more of which are cross primers. A CPA mechanism requiring two cross primers (double crossing CPA) has been previously used to detect MTB in clinical samples.

Sample preparation: use a syringe and one membrane unit, without centrifugation.
Amplification: the technology depends on the CPA provider, a water bath is the only needed instrument.
Lateral flow band detection: visual readout in a closed, cross-contamination-proof device.
Glass transition reagents: the entire kit can be transported/ stored at room temperature.

Benefits: The amplification method (CPA) and the cross-contamination-proof detection device are the primary inventions. The glass transition method and the device for sample preparation are the improvements on existing technology.
Cost-effectiveness: Installation at no cost and the instrument has a low cost.
Ease of use and maintenance: Individual testing kit, easy to use.
Low training requirements: Highly trained personnel is not required.

Steps to Follow

Step 1: Sample preparation—Use the nucleic acid extraction device in the instrument. The procedure takes 15 min once the sputum sample has been directed and has boiled.

Step 2: Amplification—Amplification can be performed in any incubator that maintains a constant temperature. The CPA takes 60 min at 63 °C.

Step 3: Detection and reading—Place the CPA reaction tube in the cartridge and close. Read after 10 min.

CPA generates amplification products visible with only four bacterial cells but does not produce a measurable amount of spurious products in the absence of a template.

The CPA method has great advantages for the detection of MTB in sputum clinical samples, especially in AFB-negative patients. This rapid detection method does not require costly equipment.

4.3.9   Smart Amplification Process (SmartAmp)

SmartAmp is a particularly useful method for single nucleotide polymorphism (SNP) and mutation detection. The method allows the detection of genetic polymorphisms or somatic mutations in ~30 min in whole blood and without the need for DNA purification.

SmartAmp uses asymmetric initiators, a turn-back (TP) initiator, and a folding (FP) initiator, that play a key role in the acceleration (TP) and control (FP) of the reaction.

4.3.10   Strand Displacement Amplification

This method was first described in 1992 and relies on bifunctional primers incorporating both target recognition and endonuclease target regions. Following strand separation, these bifunctional primers extend, incorporating the restriction target into the amplicon.

The complex asynchronous reactions occur concurrently and user interventions are limited to an initial heat denaturation with primers, followed by the addition of polymerase and restriction enzymes at a 37 °C incubation temperature. This protocol is by no means complex and appears amenable to POCT use. However, there is little mention of SDA point-of-care devices in the literature. This may be due to the original 2-h amplification process being considered as too lengthy for a POCT or to the SDA reaction's sensitivity to background DNA, that can co-amplify following nonspecific primer binding, possibly as a result of the decreased stringency conditions present at 37 °C.

An SDA method was developed (Becton Dickinson Microbiology Systems, Sparks, Md.), for the detection of mycobacteria. SDA is a method in which DNA segments of the IS6110 insertion sequence specific to MTB and a segment of the 16S rRNA gene common to all members of the genus Mycobacterium are isothermically amplified. The BDProbeTec®-SDA (Becton Dickinson) system that uses this method can automatically perform the rapid detection of mycobacteria.

This system is semi-automated and can process up to 48 samples in a single run, in under 6 h (decontamination, amplification, and detection). Due to its high sensitivity and specificity, the system can be recommended for use in clinical laboratories for the rapid detection of mycobacteria in sputum samples.

4.3.11   Nicking Enzyme Amplification Reaction

The two-stage nicking enzyme amplification reaction (NEAR) is similar to the SDA reaction, making use of nicking enzymes to generate a site from which polymerase elongation can initiate. In contrast to the SDA, the nicking enzyme employed in NEAR will only nick a single side of a duplex, removing the need (as in SDA) for strand modification of the duplex to prevent double-stranded cleavage. For POCT application, the thermal denaturation of dsDNA prior to primer binding is not required, as the primers can bind during the normal branching of DNA molecules or via nicks generated in the target genome. This reduces overall process complexity.

Once the NEAR amplification duplex is formed, nicking enzymes act on their nicking site between the promoter region and the target on both the sense and antisense strands. This divides the duplex into two parts, each with single-stranded 3′ overhangs identical to the target and a short double-stranded section comprising the nicking enzyme promoter region.

In summary, NEAR is a promising technique with a significant development pipeline currently under way. The reported performance of the technique appears to be competitive with other isothermal techniques and PCR, in terms of simplicity, sensitivity, and specificity and a clear leader in result turnover time.

4.3.12   Nicking Enzyme-Mediated Amplification (NEMA)

An engineered nicking nuclease with only three-base recognition sequence is used in combination with a DNA polymerase for the amplification of genomic DNA.

4.3.13   Isothermal Chain Amplification

In isothermal chain amplification which relies on the strand displacement activity of DNA polymerase and the RNA degrading activity of RNase H, two displacement events occur in the presence of four specially designed primers. This phenomenon leads to powerful amplification of target DNA. Since the amplification is initiated only after hybridization of the four primers, the ICA method leads to high specificity for the target sequence.

4.3.14   Exponential Amplification Reaction

An isothermal amplification method efficiently amplifies short oligonucleotides at 55 °C. The short trigger oligonucleotides that initiate exponential amplification reaction (EXPAR), called trigger X, can be enzymatically generated from specific sites within the targeted genomic DNA, and therefore represent the analyte. It has been used to detect MTB genomic DNA.

4.3.15   Limitations of Amplification Tests

Although amplification assays are sensitive and rapid in comparison to culture, there are several limitations to the first-generation amplification assays that have limited their acceptance in clinical microbiology laboratories. One of the most significant issues has been the threat of carry-over contamination, which can lead to false positives. In addition, it is difficult or impossible to verify positive amplification results in the clinical laboratory.

Another concern with first-generation assays is inhibition. All amplification assays are susceptible to substances in certain samples that can inhibit the enzymes that drive the amplification reactions.

4.3.16   Future Perspectives

Although isothermal DNA amplification can eliminate the need for a thermal cycler in POCT, the development of simple devices such as microfluidic chips is still ongoing. Such devices are integrated chips that can combine sample preparation, signal amplification and detection, thus reducing user interaction, time, and cost of analysis.

In view of the advantages of RNA amplification, a rapid test, simple to operate and with easy detection, NASBA has potential applications in clinical diagnostics and in infectious disease surveillance in developing countries, without the need for sophisticated equipment or expert personnel.

4.4   Conclusions

An overview of NAAT in TB reveals that a variety of options are commercially available or in their final development stages. These products are destined to be used in a variety of settings, from a reference laboratory to a peripheral laboratory. A great variety of NAATs are expected as an alternative to the GeneXpert in the next

Table 4.1 Comparison of various isothermal amplification methods with PCR

Property	PCR	NASBA	SMART	SDA	RCA	LAMP	HDA
DNA amplification	+	+	+	+	+	+	+
RNA amplification	+	+	+	+	+	+	+
Temperature (s) °C	94, 55–60, 72	37–42	41	37	37	60–65	R[a],37, 60–65
Number of enzymes	1	2–3	2–3	2	1	1	2
Initiator design	Sim	Sim	Com	Com	Sim	Com	Sim
Multiplex	+	+	−	−	+	−	+
Product detection	GE, ELISA, RT	GE, ELISA, RT, ECL	ELOSA, RT	GE, RT	GE, RT	GE, Turbidity, RT	GE, ELISA, RT
Inhibition tolerance	−	−			−	+	+
Template denaturing	+	+	+	+	−	−	−
Denat. agent	Heat	RNase H	RNase H	Res enz	DNA pol	Betaine	Helicase

[a]*R* room 22–24 °C, *Sim* simple, *Com* complex, *Res* enz restriction enzymes, *DNA pol* DNA polymerase, *ELISA* enzyme-linked immunosorbent assay, *ELOSA* enzyme-linked oligosorbent assay, *ECL* electrochemiluminesence, *GE* gel electrophoresis, *RT* real-time

few years. Several products offer greater processing capacity per lot, but they also require greater user intervention.

In terms of use, sample preparation is not integrated into recent NAATs, it is manually performed and this might prove to be a challenge in peripheral microscopy centers in countries with limited resources. While the cost per test is lower or similar to the GeneXpert, there are additional costs that TB programs must cover for training, implementation, and quality control.

The better known isothermal technologies for DNA or RNA amplification offer several advantages over PCR by eliminating the need for an expensive thermal cycler. However, these isothermal technologies have advantages and weaknesses that limit their use in molecular biology (Table 4.1).

The NASBA method is an innovative and powerful gene amplification technique, currently emerging as a simple and rapid diagnostic tool for the early detection and identification of infectious diseases. Using NASBA may contribute to the development of a pathogen detection kit, at an affordable price, portable, easy to use, and convenient even in the least equipped laboratories.

With the increasing need to rapidly obtain a precise DST in MDR-TB endemic regions, there is a greater number of products that include at least rifampicin resistance detection; however, most have not been endorsed and many are located in reference laboratories. Some of the new NAATs for use in peripheral laboratories have incorporated limited drug resistance tests, in integrated formats or through follow-up testing.

The need for a low-cost test that does not rely on a sputum sample but on biomarkers is a key priority. This kind of test could be applicable in community first-contact centers, not only for the diagnosis of TB but also in individuals needing a confirmation test.

References

1. Boyle D, Pai M. UNITAID: tuberculosis diagnostics technology and market landscape. 2nd ed. Geneva: WHO; 2013.
2. Fakruddin M, Mazumdar RM, Chowdhury A, Mannan KS. Nucleic acid sequence based amplification (NASBA)-prospects and applications. Int J Life Sci Pharm Res. 2012;2:L106–21.
3. Hopewell PC, Fair EL, Uplekar M. Updating the International Standards for Tuberculosis Care. Entering the era of molecular diagnostics. Ann Am Thorac Soc. 2014;11:277–85. doi:10.1513/AnnalsATS.201401-004AR.
4. Kalyan Kumar CH, Mahesh K, Mohan Reddy N. Use of loop-mediated isothermal amplification of DNA for the rapid detection of HIV/AIDS related opportunistic infections (CMV &TB) in clinical specimens. J AIDS Clin Res. 2012;3:154.
5. McCalla SE, Ong C, Sarma A, Opal SM, Artenstein AW, Tripathi A. A simple method for amplifying RNA targets (SMART). J Mol Diagn. 2012;14:328–35.
6. Sharma SK, Mohan A. Tuberculosis: from an incurable scourge to a curable disease—journey over a millennium. Indian J Med Res. 2013;137:455–93.
7. Smith JH, Beals TP. Detection of nucleic acid targets using ramified rolling circle DNA amplification: a single nucleotide polymorphism assay model. PLoS One. 2013;8:e65053.
8. Steingart KR, Schiller I, Horne DJ, Pai M, Boehme CC, Dendukuri N. Xpert® MTB/RIF assay for pulmonary tuberculosis and rifampicin resistance in adults. Cochrane Database Syst Rev. 2014;1, CD009593.
9. The Gates Foundation. Public-private partnership announces immediate 40 percent cost reduction for rapid TB test. 2012. http://www.gatesfoundation.org/press-releases/Pages/public-private-partnership-40-percent-reduction-TBtest.aspx. Accessed 25 Mar 2015.
10. World Health Organization. Automated real-time nucleic acid amplification technology for rapid and simultaneous detection of tuberculosis and rifampicin resistance: Xpert MTB/RIF system for the diagnosis of pulmonary and extrapulmonary TB in adults and children: policy update WHO/HTM/TB/2013.14. Geneva: World Health Organization; 2013a.
11. World Health Organization. Global tuberculosis report 2013. WHO/HTM/TB/2013.11. Geneva: World Health Organization; 2013b.
12. World Health Organization. Policy update: automated real-time nucleic acid amplification technology for rapid and simultaneous detection of tuberculosis and rifampicin resistance: Xpert® MTB/RIF system for the diagnosis of pulmonary and extrapulmonary TB in adults and children. Geneva: WHO; 2013c. http://www.stoptb.org/wg/gli/assets/documents/WHO%20 Policy%20Statement%20on%20Xpert%20MTB-RIF%202013%20pre%20publication%20 22102013.pdf. Accessed 25 Mar 2015.
13. World Health Organization. The use of a commercial loop-mediated isothermal amplification assay (TB-LAMP) for the detection of tuberculosis. Expert group meeting report. Geneva: World Health Organization; 2013d.
14. Steingart KR, Sohn H, Schiller I, Kloda LA, Boehme CC, Pai M et al. Xpert MTB/RIF assay for pulmonary tuberculosis and rifampicin resistance in adults. Cochrane Database Syst Rev. 2013. Jan 31;1:CD009593. doi:10.1002/14651858.CD009593.pub2. Review. Update in: Cochrane Database Sys Rev. 2014;1:CD009593.

Chapter 5
TB Infection

Abstract There are currently two available methods for the detection of *M. tuberculosis* (MTB) infection in the United States. These are: (1) the tuberculin skin test (TST) and (2) the interferon-gamma release assay (IGRA). The TST demonstrates the existence of host hypersensitivity to the tuberculosis bacillus (TB) proteins, usually as a result of MTB infection although it can also be induced with the Calmette–Guérin bacillus (BCG) or infection with environmental mycobacteria. TST provokes an inflammatory reaction with a significant dermal infiltrate at the site of tuberculin inoculation. TST requires the use of 5 PPD units. The result is expressed in terms of millimeters of induration but its interpretation is complex and depends on many variables that may influence not only the size of the reaction but also the development of false negative and false positive results, especially in cases that have been previously immunized with BCG. Until recently, there were no alternatives to TST for LTBI diagnosis. But now, there is the option of a new in vitro test: IGRA. Current evidence suggests that IGRA, especially if based on RD1 antigen cocktails, will potentially be a useful diagnostic test in the clinic and in the public health field. If this potential is attained in practice, it will have to be confirmed in large, well-designed trials and after long-term follow-up.

Keywords IGRA • Immune response • Patient immunosuppressed • TB infection • TST • Tuberculin

5.1 Introduction

The effectors of the immune response, primarily cell, are the basis for diagnostic tests *M. tuberculosis* (MTB) infection.

TB infection is considered latent if there are no associated clinical, bacteriological, or imaging signs. In these cases, positivity for infection has traditionally been detected with a positive tuberculin skin test (TST). More recently, new methods to detect infection have been developed such as in vitro immunological tests performed in whole blood and known as the interferon-gamma release assay (IGRA) [8].

Although activation of the response immune system to protect MTB infection is not completely understood, it is surprising that only 10 % of individuals develop

© Springer International Publishing Switzerland 2015
G. García-Elorriaga, G. del Rey-Pineda, *Practical and Laboratory Diagnosis of Tuberculosis*, SpringerBriefs in Microbiology,
DOI 10.1007/978-3-319-20478-9_5

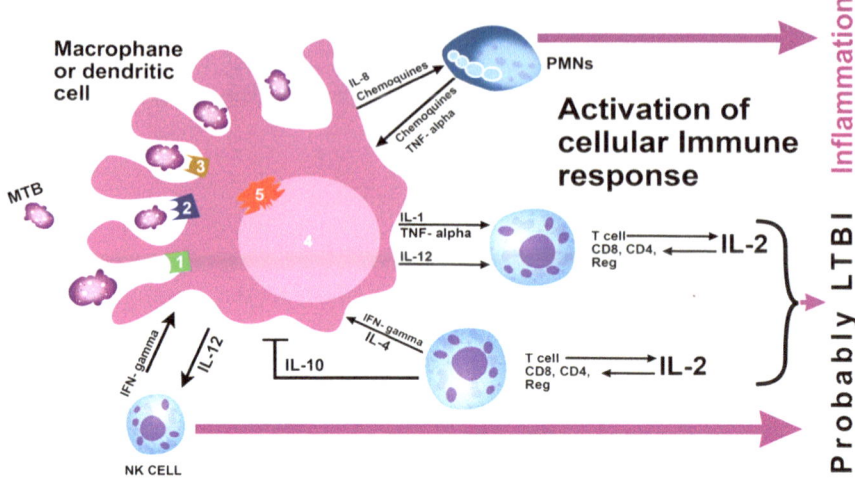

MTB.- Mycobacterium tuberculosis; 1.- Complement receptor; 2.- Mannosa receptor;
3.- Scaverger receptor; 4.- phagosome; 5.- Toll like receptor; LTBI.- Latent tuberculosis infection

Fig. 5.1 Activation of cellular immune response against *M. tuberculosis*

active disease during their lifetime, and the remainder as asymptomatic referred to as "latent" [1]. The knowledge of immunological mechanisms helps us explain why some individuals exposed to MTB infection remain latent and why others develop disease (Fig. 5.1).

The immune response to MTB has a double significance, first fight infection and the other being the cause of the disease pathology. Activation is carried out mainly by two types of mediators, cytokines and eicosanoids lipids. Both mediators are effectors and/or regulators of immune activation. Cytokines are produced by activation of genes contained within the immune cells that are activated by MTB, producing a cytokines storm such as Interleukin-12 (IL-12), tumor necrosis factor alpha (TNFα), interferon-gamma (IFNγ), also preferentially target the immune response to cellular immunity. A synthetic approach, immune activation mechanism consists of the production of IL-12 in early infection by antigen presenting cells by inducing the synthesis and secretion of IFNγ by natural killer cells and T lymphocytes, cells in turn activate macrophages to produce TNF, which promotes MTB intracellular killing by oxidative mechanisms (Fig. 5.2). Eicosanoids are enzymatically produced by way of the oxygenation of polyunsaturated omega-6 fatty acid, arachidonic acid or pathways or cyclooxygenase and lipoxygenase [10].

It is currently estimated that 11 million individuals in the United States, close to 4 % of the entire population, harbor a latent TB infection (LTBI). In most individuals, MTB infection is contained by the host's immune system and the infection remains latent. In LTBI, the bacilli that persist in asymptomatic hosts may become reactivated and lead to active disease in approximately 10 % of those infected, throughout their lifetime [8].

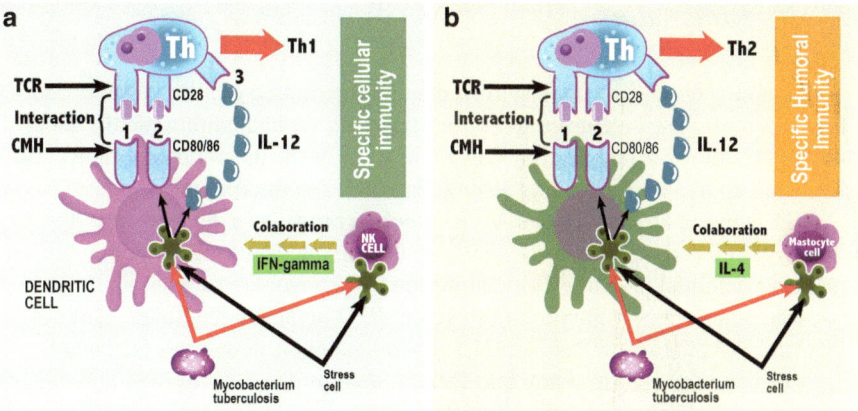

Fig. 5.2 Mediators activating the immune response against *M. tuberculosis*

We have depended on TST for decades to diagnose LTBI. The test first appeared in 1890 and it is the oldest diagnostic test in current use.

There are currently two available methods for the detection of MTB in the United States. They are:

- TST
- IGRA

QuantiFERON-TB Gold In-Tube test (QFT-GIT)
T-SPOT.*TB* test

These tests aid physicians in the differentiation of infected and non-infected individuals. However, a negative reaction to any test does not exclude the diagnosis of a latent infection or TB disease [8].

5.2 Tuberculin Skin Test

The TST is currently the only widely used method to identify MTB infection in individuals with no TB disease (TB). Although currently available TST antigens are less than 100 % sensitive and specific in the detection of MTB infection, there is no better and widely available diagnostic method.

After the initial MTB infection, it takes 2–8 weeks for the immune system to react to the purified protein derivative (PPD) and for the infection to be detected by TST. In some individuals infected with MTB, the ability of the TST to react to PPD may decrease as years go by.

5.2.1   Tuberculin

Tuberculin is obtained from a sterilized and concentrated MTB culture filtrate. The first tuberculin used (referred to as "old tuberculin") had culture media impurities and those due to mycobacterial development as well; its composition also varied from one lot to another, leading to problems in result interpretation.

TST is based on the fact that MTB infection produces a delayed-type hypersensitivity (DTH) reaction to certain antigenic components of the microorganism and that are contained in extracts of culture filtrates known as "tuberculins." There are two manufacturing companies of tuberculin PPD in the United States: Parke–Davis Pharmacy (Aplisol) and Pasteur Mérieux-Connaught Laboratories (Tubersol).

A PPD lot (lot 49608) known as PPD-S, and produced by Seibert and Glenn in 1939, is still the international standard and the standard reference material in the United States. All PPD lots should be tested to prove equipotency to PPD-S.

The standard 5-unit tuberculin dose (TU) of PPD-S is defined as the activity of the delayed skin reaction contained in a 0.1 mg/0.1 mL dose of PPD-S. The dose of the standard commercial PPD preparation is defined as the product dose that is biologically equivalent to that contained in 5 U of PPD-S.

When the PPD is diluted in a buffered diluent, it is adsorbed in variable quantities on glass and plastics. The manufacturer adds a small amount of Tween 80 detergent to the PPD diluents to decrease this adsorption.

5.2.2   Pathogenic Basis of TST

Specifically sensitized T lymphocytes proliferate in regional lymph nodes, near the site where MTB entered the host. After 2–8 weeks, these lymphocytes enter the circulation and remain there for a prolonged period of time.

5.2.3   Immunological Bases of the Tuberculin Reaction

The reaction to intradermally injected tuberculin is a classic example of a delayed (cellular) hypersensitivity reaction. T-cells sensitized by previous infection are recruited at the skin site where they release lymphokines [2].

Typically, the reaction to tuberculin begins 5–6 h after injection, maximal induration appears after 48–72 h, and collapses over a period of days. In a few people (the elderly and those tested for the first time), the reaction may not appear for 72 h.

In Fig. 5.3, it has diagrammed the immunobiology of latent TB detection from MTB penetration, secretion of immune mediators, and the final result of the TST measurement [1].

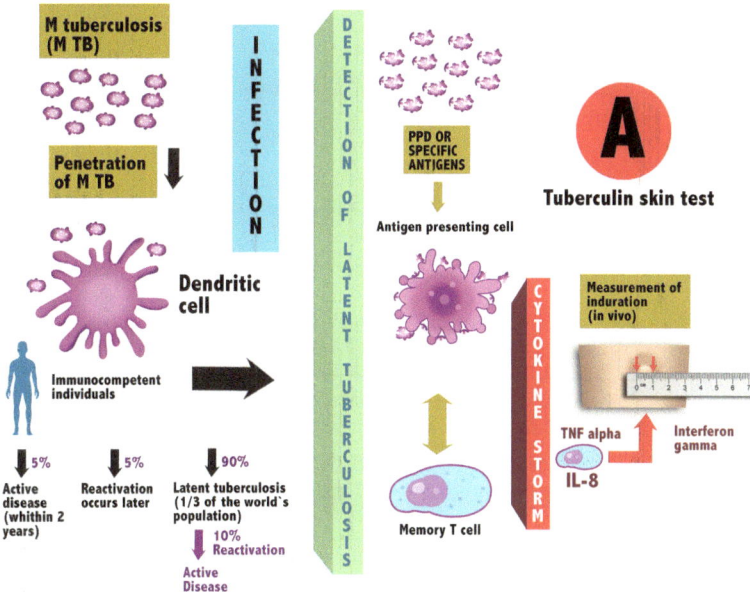

Fig. 5.3 Infection mechanism and immunobiology of *M. tuberculosis* detection test by tuberculin skin test and individuals developing active disease

5.2.4   *Factors Influencing the Test Result*

Dose

The tuberculin dose must be administered in 0.1 mL of solvent, the volume injected in the dermis. Many studies were conducted to define a PPD dose that would allow the greatest number of individuals to truly react (i.e., those infected with MTB) with the least number of false reactions.

Method of Administration

Charles Mantoux introduced and developed an intradermal method that is still common practice to date. It is recommended due to its precision and sensitivity to 5 U of PPD-S (or the bioequivalent dose of other PPD) when performing a TST in individuals known to be infected with MTB (patients with bacteriologically confirmed TB disease).

The TST, as all medical tests, is subject to variability but many of the variations inherent to its administration and reading can be avoided with careful attention to detail. The TST is performed by intradermally injecting 0.1 mL PPD containing 5 tuberculin units, on the volar or dorsal surface of the forearm. The injection must

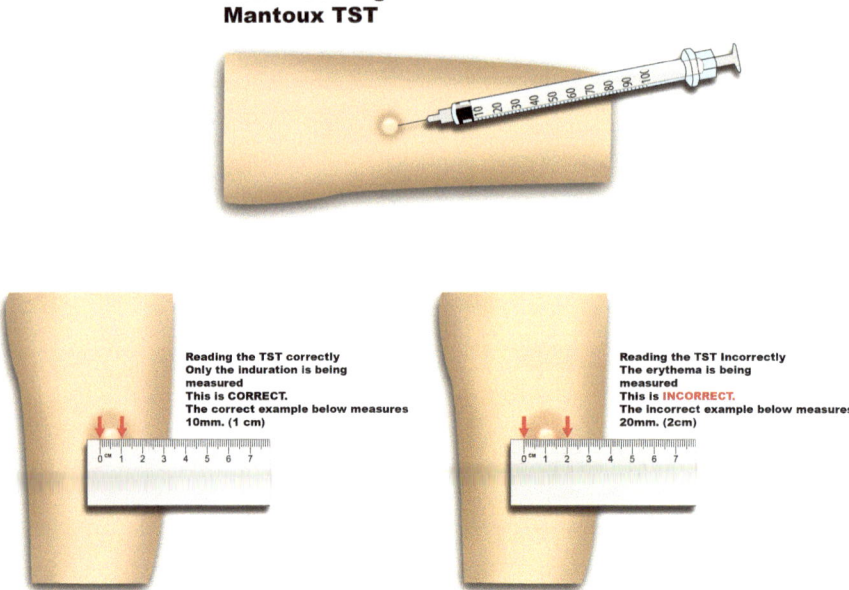

Fig. 5.4 Accurate and correct measurement in the resulting tissue reaction after administering TST

be administered intradermally with a disposable 27 caliber needle (just below the skin surface) and with the needle bevel facing upwards.

Reading Results

Before the factors influencing the test's specificity were known, test results were simply expressed as negative if there was no observable reaction or positive if it was observed. Previous attempts were made to quantify the reaction based on a cross numeration system (i.e., +, ++).

A discrete, pale elevation of the skin (hive) measuring 6–10 mm in diameter should be observed (Fig. 5.4).

Tests should be read between 48 and 72 h after the injection, when induration is maximal, by a trained health care worker. Tests read after 72 h tend to underestimate the real size of the induration.

The TST is read by palpating the injection site until an area of induration (firm swelling) is found. The diameter of the indurated area should be measured across the forearm.

Most persons can undergo a TST. TST is contraindicated only in individuals who have had a severe reaction (e.g., necrosis, blistering, anaphylactic shock, or ulcerations) to a previous TST.

With proper attention and a careful technique, the TST is very useful in measuring the prevalence of TB infection in a community. It is of limited value in diagnosing active TB (ATB) in adults because the test may be negative as a result of malnutrition or other diseases, although the patient has ATB.

Important Note

Always remember that a negative test does not exclude TB and a positive test does not indicate disease.

Additional tests required to confirm TB disease. See Chap. 3, Bacteriologic Diagnosis.

These tests help clinicians to differentiate individuals infected with MTB from those that are not. However, a negative reaction to any of these tests does not exclude the diagnosis of TB disease or latent TB infection (LTBI).

In spite of these limitations, TST is still widely used due to its ability to predict active disease in individuals infected with latent TB, and the fact that some trials have shown that treatment of LTBI, diagnosed with TST, decreases the risk of developing active disease by about 60 %.

TST Interpretation

To adequately interpret the TST, the test's sensitivity and specificity as well as its positive and negative predictive values must be thoroughly understood. A test's sensitivity refers to the percentage of individuals with infection that have a positive test.

Immune suppression may be either specific—observed in the early phases of the disease—or nonspecific as a result of drug intake, malignancy, or HIV infection. Due to the test's low sensitivity, especially in patients with acute disease and individuals infected with HIV, the TST cannot be used to eliminate the possibility of ATB. Other factors causing false negative test results are shown in Table 5.1.

Vaccination with attenuated viruses may suppress the PPD response in patients known to be infected with MTB. Live attenuated vaccines that may lead to false negative PPD results include measles, mumps, rubella, oral polio, yellow fever, BCG, and oral typhoid fever (Ty21a).

The test's specificity refers to the percentage of patients without the condition and with a negative test result. False positive results decrease a test's specificity.

In any population, the probability that a positive test reflects a true infection is influenced by the prevalence of MTB infection. Table 5.2 shows how the infection's prevalence may modify the predictive value of a positive TST (positive predictive value).

Among individuals who have been in contact with infectious TB patients, there is a 25–50 % probability of becoming infected with MTB. Likewise, in countries with high prevalence rates, adults have a high prevalence of the infection.

Table 5.1 Factors causing false negative tuberculin skin tests

Factors relating to the studied individual:
Viral, bacterial and fungal infections
Live virus vaccines
Metabolic abnormalities (chronic renal failure)
Factors relating to the type of tuberculin used:
Inadequate storage (exposure to light and/or heat)
Inadequate dilutions
Chemical denaturation
Factors relating to administration:
Injection of insufficient antigen
Subcutaneous injection
Delay in administration after loading syringe
Factors relating to test reading and result recording:
Inexperienced reader
Conscious or unconscious bias

Table 5.2 Positive predictive value of TST

Prevalence of TB infection (%)	Positive predictive value	
	Specificity of 0.95	Specificity of 0.99
90	0.99	0.999
50	0.95	0.99
25	0.86	0.97
10	0.67	0.91
5	0.50	0.83
1	0.16	0.49

Based on the sensitivity, specificity, and prevalence of TB in different groups, three cutoff points have been recommended to define a positive TST. Individuals at high risk of developing TB disease if infected with MTB, a cutoff value >5 mm is recommended.

A cutoff value >10 mm is suggested in individuals with normal or mildly compromised immunity and a high risk of being infected with MTB, but with no other risk factors that could increase their probability of developing active disease (Table 5.3). Aside from the listed groups, other high prevalence populations can be locally identified.

Individuals that most probably are not infected with MTB should not obtain a TST since the predictive value of a positive test in low prevalence populations is low. However, if a TST is performed upon entering a workplace in which a certain risk of exposure to TB is predicted and a TST program is available, a cutoff value >15 mm is recommended to improve the test's specificity (for further information, consult the ATS Supplement: Targeted tuberculin testing and treatment of latent tuberculosis infection. *Am J Respir. Crit. Care Med* 2000; 161:S221–S247). These guidelines are summarized in Table 5.4 [3].

Table 5.3 High prevalence and high risk groups

High prevalence groups	High risk groups
Individuals born in countries with high TB prevalence	Children under age 4 HIV co-infected individuals
Groups with scarce access to health care	Individuals in close contact with patients with infectious TB
Individuals that live or spend time in certain locales (i.e., shelters, prisons, housing for the underprivileged)	

Table 5.4 Classification of the Tuberculin Skin Test Reaction

Induration ≥5 mm is considered + in	Induration ≥10 mm is considered + in	Induration ≥15 mm is considered + in
HIV-infected persons	Recent immigrants (<5 years) from high-prevalence countries	Any person, including those with no known risk factors for TB
A recent contact of a person with TB	Intravenous drug users	
Persons with fibrotic changes in chest radiograph consistent with prior TB	Residents and employees in high-risk conglomerate settings	

Tuberculin Storage

Tuberculin should be stored at 4–8 °C, since its activity lessens over time outside this temperature range. This consideration is very important, especially in low-income countries that tend to have a tropical climate.

5.2.5 False Negative Readings

Some individuals have a negative TST reaction in spite of being infected with MTB.

Twenty-five percentage of individuals infected with MTB may have a negative TST at diagnosis. In many cases, those with ATB and a negative test result (anergy) have atypical TB, although a negative result is more frequent in cases of severe or disseminated disease.

Other causes of false negative TST relate to the type of administered tuberculin, its storage and the technique used to perform the test. These factors are very important since it has been estimated that untrained ancillary clinical personnel use an incorrect technique and/or interpret the test result incorrectly in 75 % of cases [6].

5.2.6 False Positive Readings

TST is a valuable tool, but it is not perfect.

False positive TST readings may occur for several reasons, although the most important is interpreting it as a TB infection while in reality the infection could be due to other environmental mycobacteria or the patient had previously received the BCG vaccine. A hematoma or a small abscess at the injection site may be interpreted as induration, while it is only secondary to the injection's trauma or another infection.

5.2.7   Boosted Reaction and Serial Tuberculin Skin Testing

Before approaching this heading, we must point out that the TST does not sensitize uninfected individuals, no matter how often it is performed. In some cases, the test must be repeated if there is a risk of acquiring TB infection, as in health care workers.

In most subjects, TST sensitivity persists throughout their lifetime. However, with time, the size of the TST reaction may decrease or even disappear. If PPD was administered to infected individuals whose skin tests have decreased, the reaction to the initial test may be small or absent; however, there may be an accentuated reaction after repeating the test. This is known as the "booster effect" and it may be misinterpreted as a skin test conversion.

Two-step testing is a strategy used to decrease the probability that a boosted reaction is misinterpreted as a recent infection. Two-step testing should be used as the initial TST in people that will be periodically tested, such as health care workers.

5.2.8   Previous BCG Vaccination

Immunization with the Calmette-Guérin bacillus (BCG) is currently used in many parts of the world as a measure to prevent TB. BCG carries the name of two French investigators that developed the vaccine from an attenuated *M. bovis* strain. There is no reliable method to distinguish tuberculin reactions due to BCG vaccination or those caused by natural mycobacterial infections. It is generally prudent to consider a "positive" reaction to 5 tuberculin UT PPD, in individuals vaccinated with BCG, as an indicator of MTB infection, especially in individuals from countries with a high TB prevalence.

5.2.9   Definition of TST Conversions

It may prove difficult to determine the relevance of TST if there are two readings with different degrees of induration. Since inevitable errors exist, even in the most carefully performed tests, small increases in the reaction's size may not be significant.

In some individuals infected with NTM or that have been vaccinated with BCG, the TST may show some degree of induration. In these subjects, a "positive" conversion is defined as an increase of 10 mm in induration in subsequent responses.

5.2.10 Anergy Test in HIV-Infected Individuals

Individuals infected with HIV may have a compromised ability to react to TST due to cutaneous anergy. The anergy test refers to the evaluation of the response to antigens in the skin test and in which a cell-mediated, DTH response would be expected.

Skin tests to detect anergy are applied by intradermal injection, using Mantoux's method, but there are no classification standards of a positive response. Individuals that respond to any antigen are considered to have a relatively intact cellular immunity, while those that do not mount a response are considered "anergic."

5.2.11 TST Indications

TST, as any diagnostic test, is performed in individuals in whom the result would lead to therapeutic intervention. There are only two possible interventions: treatment of patients with ATB and preventive therapy in infected individuals at high risk of developing TB disease.

Positivity Criteria and Indications in Low- and Middle-Income Countries: Previous considerations suggest that there is more than one TST cutoff point, hinging on the risk of acquiring TB and this complicates in great measure, its application in the field; health care workers at the peripheral level must make decisions based on each individual's risk factors. Different cutoff points are only applicable in reference centers with expert personnel and not at the peripheral health care level.

5.3 IFNγ Detection (IGRA)

For the first time, a new alternative to TST appeared in the form of a new type of assay based on T-cells in vitro: the IFNγ release assay (IGRA). These assays are based on the principle that T-cells in individuals sensitized to TB antigens produce IFNγ when exposed again to mycobacterial antigens [8]. Initial research on IGRA used PPD as the stimulation antigen, and new assays use specific MTB antigens such as early secretory antigenic target 6 (ESAT6) and the 10 kDa culture filtrate protein 10 (CFP10). These proteins, encoded by genes located in the MTB genome's region of difference 1 (RD1) locus, are more specific to MTB than PPD; they share no sub-strains with BCG or most NTM (except for *Mycobacterium kansasii*, *Mycobacterium marinum*, and *Mycobacterium szulgai*).

In terms of applicability, very few studies have been conducted in highly endemic countries with a high LTBI prevalence, high BCG coverage, and generalized exposure to LTBI. Our group conducted a cross-sectional study in Mexico, in patients infected with HIV-AIDS and suspected LTBI; our findings revealed no concordance between the TST and IGRA assays. Additional studies are required to determine IGRA's true role in the diagnosis of latent infection in these patients.

In terms of the assay methods, future research should attempt to improve IGRA's sensitivity based on the RD1 and without compromising specificity; current evidence suggests that adding more specific antigens or a combination may be effective. Comparative studies are needed to determine whether specific assay formats are associated with greater precision. Long-term cohort studies would also be useful to determine whether a positive IGRA result is associated with a greater incidence of active disease among those with latent infection. These results would help solve the debate on whether IGRA can replace the TST [4].

Current findings suggest that IGRA, especially if based on RD1 antigen cocktails, has the potential of becoming a useful diagnostic tool in the clinic and in the public health field. If this can be accomplished in practice, it will have to be confirmed in large studies, well-designed trials, and long-term follow-up.

5.3.1 Immune Response to TB Infection

The recognition of MTB during the innate immune response leads to cell activation and the production of pro-inflammatory chemokines and cytokines (Fig. 5.5). These mediators recruit inflammatory cells [T and natural killer (NK) cells, and neutrophils) to the area of infection and coordinate the adaptive immune response. After alveolar macrophages phagocyte MTB, a local, nonspecific inflammatory response develops. At this point, most mediators (IL-1b, IL-12, TNF-a, IL-15, IL-18, among others) are produced by macrophages or dendritic cells, but IFNγ is secreted by NK cells, T γδ cells, and cluster of differentiation 1 (CD1)-restricted T-cells [9, 10].

In Fig. 5.6, it has diagrammed the immunobiological mechanisms that activate the immune response in IGRA from MTB penetration to macrophage, cytokine storm until measuring IFNγ [11].

5.3.2 General Recommendations for the Use of IGRA

An IGRA, as a complementary test in immunosuppressed patients and in children, should be performed when the TST is negative since it may represent a false negative result resulting from immune abnormalities or if positive, it may reflect previous BCG immunization (especially in the previous 15 years).

IGRA tests should be performed in laboratories with accredited quality control.

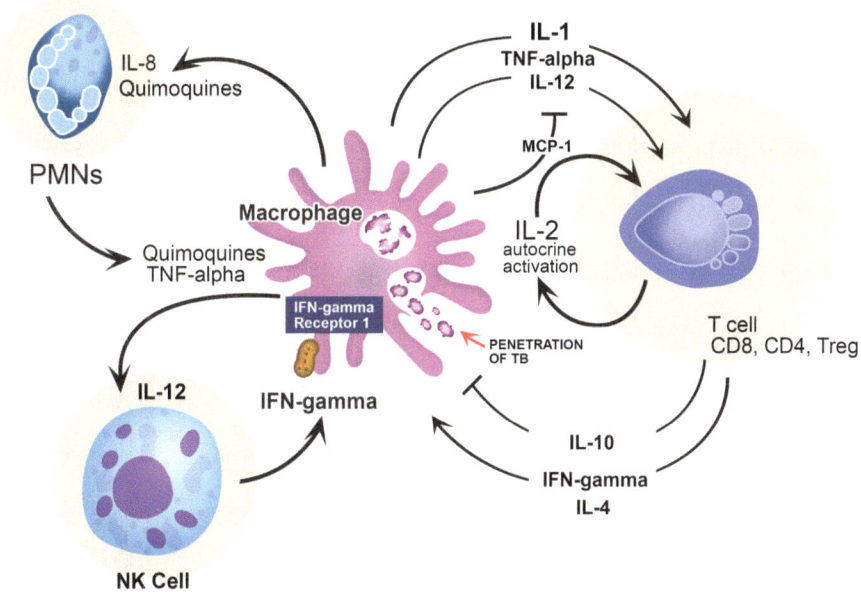

Fig. 5.5 Activation of immune response of *M. tuberculosis* infection and secretion mediators to try to contain it

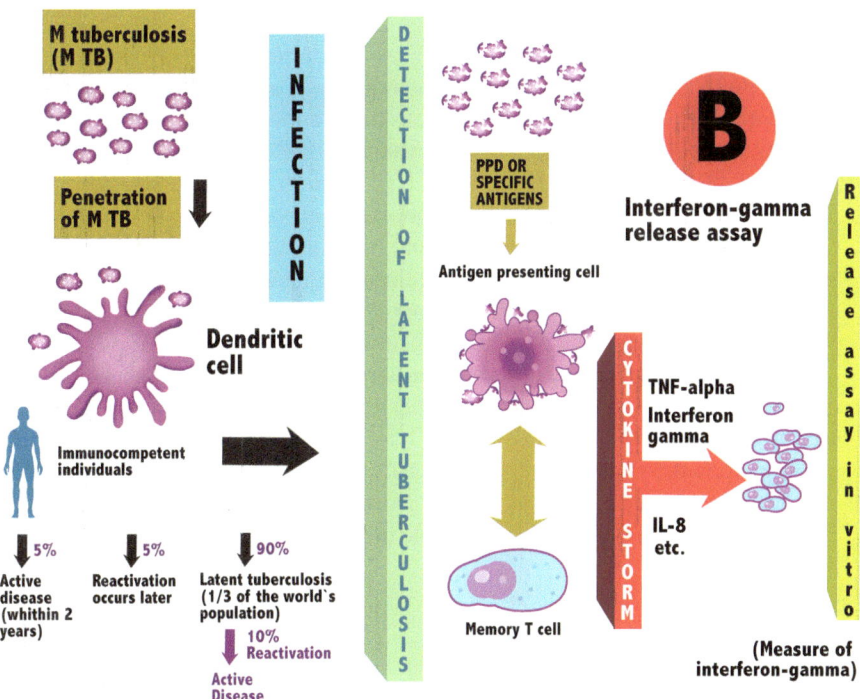

Fig. 5.6 Infection mechanism and immunobiology of *M. tuberculosis* detection test by interferon-gamma release assay and individuals developing active disease

5.3.3 Types of IGRA

Two commercial IGRA have been developed, the QuantiFERON-TB assay (Cellestis Limited, Carnegie, Victoria, Australia) and the T-SPOT.TB assay (Oxford Immunotec, Oxford, United Kingdom). Both tests measure cell-mediated immunity by measuring the interferon γ released by T-cells in response to TB antigens, following methods such as ELISA and the enzyme-linked immunospot (ELISPOT) assay. The third-generation test is now commercially available and is known as QuantiFERON-TB Gold In-Tube (QFT-GIT), and has incorporated a third mycobacterial antigen: TB 7.7, and specifically designed vials in which the blood sample is collected.

5.3.4 Test Performance and Interpretation

1. QFT-GIT. This test is performed with three specific test tubes in the reactant kit (one of the vials includes the specific TB antigens ESAT-6, CFP-10, TB 7.7— problem test tube; another contains phytohemagglutinin—positive test tube; and the third has no reactants—negative control test tube). Three milliliters of blood are required (1 mL per tube) and blood is directly withdrawn into the tubes. After shaking the vials, they are incubated for 18–22 h in a 37 °C stove, the tubes are centrifuged, and the obtained plasma is used to perform the enzymatic immuno-assay that allows the detection and quantification of the IFNγ released by the patient's lymphocytes. After incubation, the plasma may be stored for several weeks without compromising the test results.
2. T-SPOT.TB. This test requires 8–10 mL of heparinized blood. According to the manufacturer's instructions, the mononuclear cell layer must be separated and after appropriate washing and cell count, it should be adjusted to 250,000 cells/mL. This number is used as the inoculum in the four wells provided for the test (two wells have ESAT-6 and CFP-10 antigens, and the other two are the positive and negative controls). The plate is incubated for 18–22 h at 37 °C in a CO_2 stove, and the immunospot is performed, allowing the quantification of interferon-producing cells (evidenced as the number of spots; each spot represents the mark of an individual interferon-secreting T lymphocyte). An interpretation algorithm provided by the manufacturer guides result interpretation. Technically, T-SPOT.TB requires a greater blood volume, is more time-consuming, and is more labor-intensive than the QFT-GIT, and sample processing cannot be deferred.

5.3.5 Advantages of IGRA Over TST

It has been suggested that IGRA has several advantages over TST. Since the test is performed in vitro and does not require measurements such as skin induration, results are less subjective and one patient visit is sufficient (Table 5.5).

Table 5.5 TST vs. IGRA

TST	IGRA
Tuberculin is injected under the skin and causes a delayed hypersensitivity reaction if the individual has been infected with *M. tuberculosis*	Blood is extracted for testing; the test measures the immune response to TB bacteria in whole blood
Requires two or more patient visits for testing	Requires one patient visit for testing
Results available in 48–72 h	Results may be available in 24 h (depending on laboratory and transportation)

IGRA interferon-gamma release assay, *TST* tuberculin skin test

The skin test is difficult to interpret by untrained personnel; IGRA has greater specificity; the determination may be repeated immediately; results are rapidly obtained; it is easy to standardize and apply in the laboratory; better results in cases previously immunized with BCG and with cross-reactions with environmental mycobacteria; better results in young children and in the elderly; better results in immunosuppressed individuals (best studied in HIV+); protects privacy and decreases anxiety and worry over the result; avoids repeated tests (tuberculin window, booster effect) and, although more costly, this is compensated by a better cost-effectiveness ratio, a decrease in the number of work hours lost by the patient, and a significant price decrease that would entail an increase in the use of IGRAs.

We must also take into account the fact that these techniques are more expensive; they require a laboratory and personnel trained in its performance; they are not completely specific for MTB (sharing antigens with *M. kansasii*, *M. szulgai*, and *M. marinum*); they require a solid infrastructure; there may be problems after venipunctures, especially in children; results may be indeterminate and may not differentiate between a latent infection and disease, major disadvantages in and of themselves.

5.3.6 Sensitivity and Specificity

It is difficult to determine the sensitivity and specificity of these new tests in the absence of a gold standard for the diagnosis of TB infection; these new tests have been evaluated by replacing infected individuals with TB patients. In truth, it is not known whether "true positives with ATB" can be directly extrapolated to infected subjects with no disease, the group for which these tests were designed.

Three strategies have been implemented to solve the problem:

(1) evaluate patients with ATB that should be infected; (2) evaluate individuals that have been in contact with TB patients and stratify them in terms of exposure; and (3) concordance between IGRA and TST.

The specificity of these tests may be estimated in individuals previously immunized with BCG and no risk factors for TB development, thus assuming they do not have a latent infection.

Individuals with no apparent risk of infection, is not completely dependable to be used as controls. All data appear to confirm that the expected inverse ratio between sensitivity and specificity appears to be fulfilled. We must also remember that both TST and IGRA are an indirect measure of host exposure to MTB and are only ancillary tests for the diagnosis of disease and indicative or guiding in the diagnosis of infection.

Regardless of the in vitro technique under consideration, the specificity of IGRA is superior to that of TST.

There is discordance between IGRA and TST results in 10–40 % of patients. This discordance is greater in immunized than in non-immunized individuals.

Although the evidence provided by IGRA is indirect for the establishment of a latent infection diagnosis, it still has good sensitivity and in the immunized population, high specificity.

5.3.7   Clinical Performance of IGRA

Determination Techniques in Immunocompetent Patients: An important facet of TB transmission is that the risk of contagion is mainly determined by the frequency, duration, and proximity of contact with the individual diagnosed with TB. Therefore, a new diagnostic test with greater sensitivity and specificity than TST should be more closely related with the level of exposure and be independent of previous immunization (BCG).

Studies have ascertained that QFT-GIT and T-SPOT.TB results are not affected by previous BCG immunization which does occur with TST; this underscores their greater specificity. Also, IGRA correlates better with exposure to TB disease.

5.3.8   IGRA in Immunosuppressed Patients

Tuberculosis and HIV

TB has become the most important coinfection in patients infected with HIV, a situation affecting approximately 13 million individuals worldwide. In Africa, it is the primary cause of death in HIV-infected patients and it is also the most frequent disease in AIDS patients on antiretroviral therapy. The diagnosis of LTBI in HIV-infected patients is traditionally based on a TST that aside from the previously referred drawbacks, adds a significant rate of anergy to these patients.

Compared to TST, data on IGRA suggest it has superior specificity, a lower number of false positive results due to previous BCG immunization, and greater sensitivity in populations with a low incidence of TB. However, there is scant data describing the performance of IGRA in individuals infected with HIV and whose immunological abnormalities could compromise the test's performance since it is based on lymphocyte activation.

In terms of IGRA's diagnostic performance in HIV-infected patients, T-SPOT.TB is more sensitive than TST and better correlates with ATB.

Indeterminate results are closely dependent on the patient's CD4 count. Thus, indeterminate results in patients with a CD4 count ≥ 100 cells/mm^3 are present in 3 % of cases, while they are present in 24 % of patients with CD4 cell counts <100 cells/mm^3.

Patients with deteriorating cellular immunity (i.e., HIV infection, immunosuppressive therapy including TNF inhibitors) are at high risk of developing the disease if they acquire the infection. While in the non-immunosuppressed population, the risk of developing TB from the moment the infection is acquired is 10 % throughout the individual's lifetime, and the risk in HIV patients is 10 % annually.

There are few studies evaluating the usefulness of these tests in immunosuppressed populations [7]. T-SPOT.TB has been considered as more sensitive than TST and QFT TB-G, with a lower percentage of indeterminate test results.

5.3.9 Cost-Effectiveness

After studying the contacts of TB patients, it is recommended in the detection of infected individuals that may subsequently develop the disease. Treatment of infected individuals, particularly with isoniazid, has been shown to decrease the future number of TB cases.

IGRA use guidelines vary from country to country. For instance, the Centers for Disease Control (CDC) in the United States recommend the replacement of TST by IGRA in all cases.

For the study of contacts, the TST/IGRA strategy using both tests is more economical than T-SPOT.TB, QFT-GIT, or TST alone. However, costs are considerably different depending on the test and the country in which it is applied.

5.3.10 International Guidelines on IGRA Use

There are currently three guidelines: those from the CDC in the United States; the NICE guidelines in the United Kingdom; and the Swiss that recommend the incorporation of IGRA in the study of contacts.

In May 2005, the CDC approved the use of QFT-GIT as an aid in the diagnosis of TB infection, including ATB and LTBI. According to these guidelines, QFT-GIT may be used instead of (and not in addition to) TST, in circumstances requiring its performance, including in the study of contacts, the evaluation of immigrants immunized with BCG, and in health care personnel that has been in contact with TB patients.

They recommend that the sanitary measures followed when a patient's QFT-GIT is positive, should be similar to those taken with a positive TST; there is no reason to perform a TST in a patient with a positive QFT-GIT.

How to follow patients with an indeterminate QFT-GIT remains unknown. Possible options include: repeat the test, perform a TST, or do nothing.

British norms on the use of these tests are more restrictive than the American ones. Their diagnostic strategy takes into account the cost-effectiveness ratio when studying contacts and recommends performing a TST first; if positive or if results are not very dependable, IGRA should follow, if available.

References

1. Achkar JM, Chan J, Casadevall A. B cells and antibodies in the defense against *Mycobacterium tuberculosis* infection. Immunol Rev. 2015;264:167–81.
2. American Thoracic Society. Diagnostic standards and classification of tuberculosis in adults and children. This official statement of the American Thoracic Society and the Centers for Disease Control and Prevention was adopted by the ATS Board of Directors. July 1999. This statement was endorsed by the Council of the Infectious Disease of America, September 1999. Am J Respir Crit Care Med. 2000a;161:1376–95.
3. American Thoracic Society. Targeted tuberculin testing and treatment of latent tuberculosis infection. This official statement of the American Thoracic Society was adopted by the Board of Directors, July 1999. This is a joint statement of the American Thoracic Society (ATS) and the Centers for Disease Control and Prevention (CDC). This statement was endorsed by the Council of the Infectious Diseases Society of America (IDSA), September 1999, and the sections of this statement as it relates to infants and children were endorsed by the American Academy of Pediatrics (AAP), August 1999. Am J Respir Crit Care Med. 2000b;161(4 Pt 2):221–47.
4. Arias Guillén M. Advances in the diagnosis of tuberculosis infection. Arch Bronconeumol. 2011;47:521–30.
5. Cliff JM, Kaufmann SHE, McShane H, Van Helden P, O'Garra A. The human immune response to tuberculosis and its treatment: a view from the blood. Immunol Rev. 2015;264:88–102.
6. Craviotto FG, Limongi L. Controversias en el uso del derivado proteico purificado de tuberculina (PPD) y las nuevas técnicas en la detección in vitro de los niveles de interferón gamma (IGRAs) en un país con alta tasa de infección por tuberculosis. Rev Am Med Respir. 2012;2:44–53.
7. García-Elorriaga G, Martínez-Velázquez M, Gaona-Flores V, Del Rey-Pineda G, González-Bonilla C. Interferon γ in patients with HIV/AIDS and suspicion or latent tuberculosis infection. Asian Pac J Trop Med. 2013;6:135–8. doi:10.1016/S1995-7645(13)60009-7.
8. Jasenosky LD, Scriba TJ, Hanekom WA, Goldfeld AE. T cells and adaptive immunity to *Mycobacterium tuberculosis* in humans. Immunol Rev. 2015;264:74–87.
9. Majlessi L, Prados-Rosales R, Casadevall A, Brosch R. Release of mycobacterial antigens. Immunol Rev. 2015;264:25–45.
10. Mayer-Barber KD. Cytokine and lipid mediator networks in tuberculosis. Immunol Rev. 2015;264:264–75.
11. Nathan C. What can immunology contribute to the control of the world's leading cause of death from bacterial infection? Immunol Rev. 2015;264:2–5.

Index

© Springer International Publishing Switzerland 2015
G. García-Elorriaga, G. del Rey-Pineda, *Practical and Laboratory*
Diagnosis of Tuberculosis, SpringerBriefs in Microbiology,
DOI 10.1007/978-3-319-20478-9